TORNADOES

GREAT DISASTERS

TORNADOES

Nancy Harris, *Book Editor*

Daniel Leone, *President*
Bonnie Szumski, *Publisher*
Scott Barbour, *Managing Editor*

**GREENHAVEN
PRESS** ®

San Diego • Detroit • New York • San Francisco • Cleveland
New Haven, Conn. • Waterville, Maine • London • Munich

© 2003 by Greenhaven Press. Greenhaven Press is an imprint of The Gale Group, Inc., a division of Thomson Learning, Inc.

Greenhaven® and Thomson Learning™ are trademarks used herein under license.

For more information, contact
Greenhaven Press
27500 Drake Rd.
Farmington Hills, MI 48331-3535
Or you can visit our Internet site at http://www.gale.com

Cover credit: © Gerhard Steiner/CORBIS

Andrea Booher/FEMA News Photo, 11
National Oceanic and Atmospheric Administration, Central Library, National Severe Storms Laboratory, 21, 38, 63, 87
National Oceanic and Atmospheric Administration, Historic National Weather Service Collection, 48, 76

LIBRARY OF CONGRESS CATALOGING-IN-PUBLICATION DATA
Tornadoes / Nancy Harris, book editor.
p. cm. — (Great disasters)
Includes bibliographical references and index.
ISBN 0-7377-1473-5 (pbk. : alk. paper) — ISBN 0-7377-1472-7 (lib. : alk. paper)
1. Tornadoes. I. Harris, Nancy. II. Great disasters (Greenhaven Press)
QC955 .T68 2003
363.34'923—dc21   2002072764

Printed in the United States of America

CONTENTS

Chapter 1: The Science and Study of Tornadoes

1. The Nature of Tornadoes

A research meteorologist describes what a tornado is,
what it looks and sounds like, and some of its
destructive characteristics.

2. Exploring the Mysteries of Tornadoes

In the spring of 1994 and 1995, a crew of scientists
studied tornadoes in the field not only to collect data
on how tornadoes form but also to help improve
forecasting. The Verification of the Origin of Rotation in Tornadoes Experiment (VORTEX) was an
intense project involving more than seventy-five
researchers.

3. Storm Chasers

A lead weather forecaster and a research meteorologist lead a group in a day of tornado chasing. After
searching most of the day, the chasers witness the
brief touchdown of a translucent twister.

4. Measuring Tornadoes with TOTO

The Totable Tornado Observatory (TOTO), a tornado measuring instrument, was field-tested in the
early 1980s. The four-hundred-pound device was
placed in the probable paths of tornadoes to determine the wind, pressure, temperature, and electric
fields of tornadoes.

Chapter 2: Disasters and Personal Accounts

Chapter 3: Averting Disaster

H umans have an ambivalent relationship with their home planet, nurtured on the one hand by Earth's bounty but devastated on the other hand by its catastrophic natural disasters. While these events are the results of the natural processes of Earth, their consequences for humans frequently include the disastrous destruction of lives and property. For example, when the volcanic island of Krakatau exploded in 1883, the eruption generated vast seismic sea waves called tsunamis that killed about thirty-six thousand people in Indonesia. In a single twenty-four-hour period in the United States in 1974, at least 148 tornadoes carved paths of death and destruction across thirteen states. In 1976, an earthquake completely destroyed the industrial city of Tangshan, China, killing more than 250,000 residents.

Some natural disasters have gone beyond relatively localized destruction to completely alter the course of human history. Archaeological evidence suggests that one of the greatest natural disasters in world history happened in A.D. 535, when an Indonesian "supervolcano" exploded near the same site where Krakatau arose later. The dust and debris from this gigantic eruption blocked the light and heat of the sun for eighteen months, radically altering weather patterns around the world and causing crop failure in Asia and the Middle East. Rodent populations increased with the weather changes, causing an epidemic of bubonic plague that decimated entire populations in Africa and Europe. The most powerful volcanic eruption in recorded human history also happened in Indonesia. When the volcano Tambora erupted in 1815, it ejected an estimated 1.7 million tons of debris in an explosion that was heard more than a thousand miles away and that continued to rumble for three months. Atmospheric dust from the eruption blocked much of the sun's heat, producing what was called "the year without summer" and creating worldwide climatic havoc, starvation, and disease.

As these examples illustrate, natural disasters can have as much impact on human societies as the bloodiest wars and most chaotic political revolutions. Therefore, they are as worthy of study as the major events of world history. As with the study of social and political events, the exploration of natural disasters can illuminate the causes of these catastrophes and target the lessons learned about how to mitigate and prevent the loss of life when disaster strikes again. By examining these events and the forces behind them, the Greenhaven Press Great Disasters series is designed to help students better understand such cataclysmic events. Each anthology in the series focuses on a specific type of natural disaster or a particular disastrous event in history. An introductory essay provides a general overview of the subject of the anthology, placing natural disasters in historical and scientific context. The essays that follow, written by specialists in the field, researchers, journalists, witnesses, and scientists, explore the science and nature of natural disasters, describing particular disasters in detail and discussing related issues, such as predicting, averting, or managing disasters. To aid the reader in choosing appropriate material, each essay is preceded by a concise summary of its content and biographical information about its author.

In addition, each volume contains extensive material to help the student researcher. An annotated table of contents and a comprehensive index help readers quickly locate particular subjects of interest. To guide students in further research, each volume features an extensive bibliography including books, periodicals, and related Internet websites. Finally, appendixes provide glossaries of terms, tables of measurements, chronological charts of major disasters, and related materials. With its many useful features, the Greenhaven Press Great Disasters series offers students a fascinating and awe-inspiring look at the deadly power of Earth's natural forces and their catastrophic impact on humans.

The largest and deadliest tornado in U.S. history was the Tri-State tornado of March 18, 1925. It traveled along a 219-mile path through Missouri, Illinois, and Indiana. The tornado exhibited several deadly characteristics. Its speed was estimated at sixty miles per hour, which is twice the typical forward speed of a tornado. In addition, no visible funnel cloud was seen, and, although the tornado crossed mountains, plains, and rivers and smashed through town after town, the twister never lifted or skipped off the ground but proceeded on a straight path of destruction. According to a 1925 St. Louis newspaper account, "Boards, poles, cans, garments, stoves, whole sides of little frame houses, in some cases the houses themselves, were picked up and smashed to earth. And living beings, too. A baby was blown from its mother's arms. A cow, picked up by the wind, was hurled into the village restaurant." In three and a half hours, the Tri-State tornado killed almost 700 people, injured 2,000, and left more than 10,000 homeless.

Another memorable disaster happened in the spring of 1974, when the largest outbreak of tornadoes ever recorded, the "Super Outbreak," swept through thirteen states. In less than twenty-four hours, 148 twisters left a path of destuction twenty-five hundred miles long, demolishing the towns of Xenia, Ohio, and Brandenburg, Kentucky. At one point, as many as five large tornadoes were on the ground simultaneously, and in the state of Kentucky, it was recorded that no fewer than twenty-six twisters touched down during the outbreak. Scientists took photos from the air that revealed paths of tornadoes that had climbed three-thousand-foot ridges and plunged across deep mountain valleys. On this day, called "Terrible Tuesday," tornadoes killed an estimated 315 people, injured more than 6,000, and destroyed almost 10,000 homes.

Tornadoes are composed of the most powerful winds on earth and can destroy entire towns in a matter of minutes. In addition to the intense winds of a tornado, rapid spinning adds to its destructive fury by creating a vacuum within its funnel. This vacuum can suck up cars, trees, homes, and sometimes people, carrying them as much as a mile away.

Many tornadoes are spawned from large thunderstorms and are accompanied by thunder, lightning, rain, and hail. However, a large variety of weather patterns can lead to tornadoes, and often, similar patterns may produce no severe weather at all. Because of this, as well as the fact that tornadoes leave behind no craters, gas emissions, or faults, tornadoes are difficult to study and predict. In addition, most tornadoes come and go in a matter of minutes and can barely be seen because they are so weak. Although most tornadoes in the United States move from the southwest to the northeast, it is impossible to consistently predict where they will head, and some have been known to completely reverse their direction. They can travel in straight or erratic paths and seem to randomly destroy as they go—a house here, a house there.

Scientists are still unsure exactly how tornadoes form. However, during the last few decades, they have gathered tremendous amounts of information and have improved their ability to predict tornadoes and prevent their deadly consequences.

Systematic attempts to predict tornadoes began in Oklahoma in the late 1940s after a tornado hit an air force base. Scientists have since collected voluminous amounts of data in lengthy field

The rapid spinning of a tornado creates a powerful vacuum that can suck up cars, trees, homes, and even people.

studies to unravel the mystery of tornadoes. "Storm chasers," scientists who track storms and record data, have helped by collecting photos and making observations. Some researchers and forecasters spend every spring vacation hunting for twisters. These chasers travel hundreds of miles a day but may see only one or two serious tornadoes per season. Some have accumulated as many as ten thousand miles in the field.

Technological advances have also helped scientists to better understand and predict tornadoes. Scientists have begun using satellites that can detect developing storms and improved Doppler radar to measure the speed and direction of winds in a thunderstorm. These measurements help to determine if the storm environment is ripening for the formation of a tornado. Researchers are also attempting to use computers to improve the lead (warning) time for tornadoes. The goal is to develop forecasting tools that will allow the National Weather Service to pinpoint where the most intense part of a storm will hit six hours before it arrives. In addition, researchers hope to predict the type, location, and intensity of new thunderstorms up to two hours before they form. Government officials and meteorologists have encouraged the use of weather radios for the public to get the most up-to-date weather forecasts and warnings.

Along with their efforts to understand and predict tornadoes, experts are attempting to better educate the public about twisters. Government programs have been implemented in communities to inform the public about tornado preparedness. Programs have also provided assistance in times of tornado disaster. These efforts have paid off in the form of lives saved. In the 1950s, the yearly tornado death toll was in the hundreds, but current figures are in the tens per year.

In spite of their destructiveness, tornadoes serve important purposes in the earth's weather patterns. A tornado, like a whistle on a teakettle, helps let off pressure that builds up in the atmosphere. In addition, the storms that produce tornadoes often bring much-needed rain. Humankind must learn to live safely with tornadoes.

Chapter 1

The Science and Study of Tornadoes

The Nature
of Tornadoes

BY THOMAS P. GRAZULIS

*In the following selection, meteorologist Thomas P. Grazulis describes
the tornado as a "wonder of nature" capable of inciting both awe and
dread. He explains that a tornado is a vortex produced by a thunderstorm in which air is moving quickly in an upward spiral, often carrying
dust and debris along with it. Although they are rare, and although 80
percent of them are weak and leave no damage, tornadoes can be extremely destructive, with winds reaching up to 250 or even 300 miles per
hour. The most powerful tornadoes can move cars half a mile, cross mountains, and carry people as far as a mile from their homes. The typical tornado is only about fifty yards wide and travels about one mile. However,
some grow to be a half mile wide and may stay on the ground for an
hour or more, leaving a trail of destruction forty miles long.*

*Thomas P. Grazulis is a tornado research meteorologist and the
founder and director of the Tornado Project, a small company that gathers
and compiles tornado information.*

Few other phenomena can form and vanish so quickly, leave
behind such misery, and still be seen as beautiful. Tornadoes
consist of little more than a mixture of insubstantial air and
water vapor. Every year a few thousand people are witness to one
of the most spectacular natural forces on this planet. Seemingly
out of nothing, an enormous apparition takes shape in the sky.
Within seconds it begins to evoke awe, curiosity, inspiration, in-
significance, helplessness, even denial, and, for many, a closeness
to the power of the Creator that they have never felt before. Unfortunately, this wonder of nature does not just inspire and challenge the senses, it randomly wreaks havoc on people's lives.

It is not difficult to understand why someone might be in-

Thomas P. Grazulis, *The Tornado: Nature's Ultimate Windstorm*. Norman: University
of Oklahoma Press, 2001. Copyright © 2001 by University of Oklahoma Press.
Reproduced by permission.

trigued by an enormous funnel that seems to chew its way through a town. Something larger than the Empire State Building should not be able to move at all, much less gracefully. Some people dream about tornadoes. Others consider tornadoes to be the direct wrath of God. Many people see them as the perfect earthly link to the forces of the cosmos—gigantic in size, displaying immense power but not as unfathomable as a supernova or so rare as a planet-asteroid collision.

Difficult to Study

Tornadoes are very difficult to study. After a flood one can examine a river and its watershed. After a volcanic eruption one can study a crater and exhaust gases. After an earthquake one can survey faults. But after a tornado has "roped out" and the thunderstorm dissipated, we are left with nothing more than piles of debris and more questions.

The most intense tornadoes are born in the turbulence of a thunderstorm, arguably one of the most hostile environments on earth. The thunderstorm protects the tornado from researchers with 100-mph straight winds, baseball-sized hail, 30,000,000-volt lightning discharges, and blinding rain. After the tornado dissipates the problem of documentation arises. This is expensive and time-consuming work, often requiring detailed aerial and ground surveys. Most of the limited resources of the United States are put first into forecasting and then into rescue and cleanup. Very little money is devoted to tornado documentation.

Tornadoes do not just enter a home unexpectedly. They rip it apart and scatter the accumulated possessions of a lifetime to the four winds. They present a much different threat than do floods, lava flows, and earthquakes. Rivers, volcanoes, and faults are ever-present signatures of a risk that some people willingly accept. Automobiles, swimming pools, and railroad crossings are vastly greater threats to human life than are tornadoes. They take 40,000, 4,000, and 1,000 lives a year, respectively, compared to only about 80 for tornadoes. The difference is that these other risks are obvious, and we accept them as part of an improved lifestyle. For many people, that acceptance of risk does not extend to tornadoes. The potential loss of all one's possessions in the blink of an eye is hard to live with. It is this aspect of tornadoes, along with their visible display of power, that is the source of so much inappropriate worry.

Despite the frequency of newspaper stories about them and
the availability of images on videotape, tornadoes are uncommon.
Only about one in a thousand thunderstorms produces a tornado.
Most tornadoes (more than 80 percent) are weak and inflict no
more damage than would straight-line winds in a severe thunderstorm. Less than 1 percent of the American population will
ever be in the path of even the weakest tornado during their lives.
Even in the most tornado-prone areas of the country, a home can
expect to be hit only about once in a thousand years. The frequency of actual destruction of any given house in the heart of

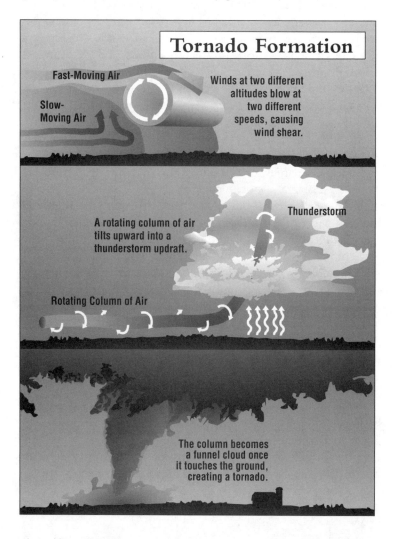

Tornado Formation

Fast-Moving Air

Slow-
Moving Air

Winds at two different
altitudes blow at
two different
speeds, causing
wind shear.

A rotating column of air
tilts upward into a
thunderstorm updraft.

Thunderstorm

Rotating Column of Air

The column becomes
a funnel cloud once
it touches the ground,
creating a tornado.

tornado alley is only about once in 10,000 to 1,000,000 years, depending largely on one's definition of the word *destruction*.

While an unlikely part of most of our lives, tornadoes are a force that cannot be ignored. The most intense tornadoes can carry automobiles half a mile and level a well-built home. Some have crossed mountains, seemingly unimpeded. Tornadoes have carried people as far as a mile from their homes. They have lasted for more than an hour while scouring the earth with wind speeds of 250 mph. Tornadoes have occurred in every state, and each area of the country has its own unique "tornado season." In a few states they have occurred in every month and at all times of day and night. Tornadoes are not unique to the United States. They have killed up to one thousand people in a single swath across Bangladesh, for example.

Tornado Characteristics

A tornado is a comparatively organized structure. In its simplest form it is a single vortex in which air, often laden with dust and debris, is moving at very high speeds in an upward spiral. The rising air enters the vortex at its base and exits in the upper part of the funnel. Most of the debris is centrifuged out very quickly, in the bottom few hundred feet. Some of the lighter debris, however, can become caught in the upward spiral and carried for several miles above the earth's surface. The vortex itself can extend 5 miles or more upward from the surface, high into the parent thunderstorm. We may see only the bottom 10 percent of it protruding from the base of the thunderstorm as a "tornado."

In a single-vortex tornado, the surrounding air rushes inward toward a central low pressure, then upward within the outer wall of the tornado. The speed of the upward moving air may be about the same as the horizontal speed or rotation. The funnel could have air rotating at 100 mph and rising at 100 mph. The combined wind speed at any given point in this spiral may vary from less than 70 mph to (very rarely) as much as 300 mph. In some tornadoes the upward component may greatly exceed the horizontal rotation, lifting buildings and causing extreme levels of destruction.

To be called a tornado, the phenomenon must be a naturally occurring atmospheric vortex whose circulation extends from the ground at least to the base of a convective cloud. Dust devils form under a clear sky and are never called tornadoes. Oil-fire, forest-fire, and prairie-fire vortices are also not classified as tornadoes, even though some may connect with an overhead cloud produced by the fire. Fire-induced vortices are not considered tornadoes, even if they look like one, travel away from the fire, and kill people, as one did in California in 1926.

The tornado funnel can take a variety of forms, depending on the immediate conditions of air pressure, temperature, moisture, dust, the rate at which air flows into the vortex, and whether the air in the core of the tornado is moving upward or downward.

Tornado Descriptions

Tornadoes have been colorfully described in many ways: a giant serpent with its head feeding on the ground; the finger of God; a huge elephant's trunk searching for food; a monstrous snake, writhing, biting, and kissing the ground; a giant barrel hanging in the air; a great column surrounded by silvery ribbons; or as ropes, balloons, beehives, or hourglasses. Some tornadoes emit only a faint, high-pitched whine. Others have been described as sounding like a waterfall, a freight train, the buzzing of a million bees, and even the bellowing of a million mad bulls. The degree of interaction with the ground may have something to do with the sound that is generated. A tornado is a very long, whirling tube of air, an enormous acoustical instrument, with a hollow core and debris-filled cone or cylinder. No one has fully explored the sound-generating properties of such an object.

The typical (median) tornado is only about 50 yards wide and has a path length of about 1 mile. However, each year a few grow to a half mile wide and stay on the ground for an hour or more, carving out a 40-mile or longer path of destruction. About one thousand tornadoes a year are counted, but twice as many may actually touch down.

Exploring the Mysteries of Tornadoes

BY PETER N. SPOTTS

In 1994 and 1995 more than seventy-five researchers from the United States and Canada were involved in a two-year study of tornadoes called VORTEX (Verification of the Origin of Rotation in Tornadoes Experiment). They hunted for twisters in the states of Texas and Kansas in order to improve the warning time as well as the accuracy of tornado predictions. New information was gathered on how tornadoes are formed and was fed into computer models to try to simulate and forecast tornadoes. The researchers were able to study ten tornadoes, including a severe twister in Dimmitt, Texas, which became the most monitored tornado in history. Despite its great wealth of information, the project failed to yield a definitive answer to the daunting question of how a tornado is triggered. Peter N. Spotts is a staff writer for the Christian Science Monitor.

The vans and compact cars looked like refugees from Mystery Science Theater 3000 as they took to the highway in search of tornadoes. Wind gauges spun, and weather vanes swung erratically from poles anchored to rooftop racks as the caravan headed out.

The scientists inside the vehicles aimed to use the instrument-laden squadron to take close-up readings that would help them pry open the secrets of how such destructive wind storms form. In the year since their final foray—part of a two-year effort known as the VORTEX project—researchers who roamed from Texas to Kansas in search of twisters have begun poring through the data they've gathered.

VORTEX, which involved more than 75 researchers from 10

Peter N. Spotts, "Twisters Still Hold Mysteries for Chasers," *The Christian Science Monitor*, vol. 88, May 23, 1996, p. 10. Copyright © 1996 by *The Christian Science Monitor*. Reproduced by permission.

universities and several federal agencies in the United States and Canada, returned a wealth of observations that participants estimate will take several years to interpret. The ultimate goal: to help save lives by improving the lead time and accuracy of tornado warnings.

North America is not the only tornado hot spot. Australia and Bangladesh also account for a large number of the world's tornadoes. [In May 1996], for example, a twister packing winds estimated at 125 miles an hour spent half an hour leveling 80 Bangladeshi villages, leaving an estimated 615 dead and 34,000 injured.

The United States has the dubious distinction of being the tornado capital of the world. [The year 1995] set a record, with 1,233 confirmed sightings in states ranging from Texas to New Hampshire and Florida to Oregon. Since 1950, no state has been spared. Indeed, Southern California experiences as many tornadoes as the Oklahoma City area, says Roger Wakimoto, a professor of meteorology at the University of California at Los Angeles. The difference, he quickly adds, is that California's twisters are much weaker and shorter-lived.

Although tornadoes can occur during any season, they become most active from late winter to early summer, starting near the Gulf of Mexico in March and moving north and east as the seasons progress. Outbreaks typically peak in May and June in Kansas, Iowa, and Nebraska.

Weaker tornadoes, with wind speeds of around 110 miles an hour, last less than 10 minutes. They cut a swath roughly 100 yards wide for about a mile before they disappear. Strong tornadoes have been known to last for up to two hours or more, cutting a path up to 1,000 yards wide and more than 100 miles long. Estimates of the maximum wind speeds in the strongest tornadoes have reached as high as 280 miles an hour.

Because they can develop suddenly and vanish just as fast, tornadoes pose a particularly thorny problem for forecasters.

What VORTEX Did

VORTEX, which stands for Verification of the Origin of Rotation in Tornadoes Experiment, was designed to test current notions of how tornadoes form, as well as provide new information that could be fed into computer models that try to simulate and forecast tornadoes.

During the springs of 1994 and '95, a small army of scientists and graduate students chased storms and tried to surround tornadoes with their armada of specially instrumented cars. They dotted the landscape ahead of tornadoes with small instrument packages called "turtles," in hopes a twister would inhale one and give researchers the inside story. Mobile Doppler radar yielded unique close-up images of storm circulation patterns, while sounding balloons and the National Oceanic and Atmospheric Administration's hurricane hunter aircraft rounded out the assault force.

In the end, the team bagged 10 tornadoes—including a severe twister that hit Dimmitt, Texas, and became the most monitored tornado in history. They also probed numerous thunderstorms with tornado potential.

"VORTEX was able to get more information about tornadoes from more points" than ever before, says Charles Doswell, a scientist at the National Severe Storms Laboratory at the University of Oklahoma in Norman. From a scientific standpoint, he says, the VORTEX data are exciting—and humbling.

Going into the project "we had some sense that we knew what was going on," he explains. Instead, "we saw enough variability in the way tornadoes form that . . . each one was unique. . . . As one of my colleagues put it, we've defined the depths of our ignorance."

Project VORTEX researchers used vehicles equipped with special instruments to gather information on how tornadoes are formed.

As daunting as the VORTEX data may be, the understanding of conditions that lead to tornado-generating thunderstorms has come a long way since Benjamin Franklin saddled up to chase dust devils. "Even before radar and traditional meteorology started after World War II, people thought that a tornado's rotation developed within a storm and then fell to the ground," explains Dr. Wakimoto, a member of the VORTEX team. After all, that's what people saw—a dark funnel cloud descending from the base of a thunderstorm.

By the mid-1970s, researchers at the University of Oklahoma were developing radar that takes advantage of the Doppler effect to detect wind circulation in clouds. Doppler radar measures the shift in frequency that results when its signals bounce back off of moving water droplets in the clouds. If part of the reflected beam rises in frequency compared with the outgoing beam, that part of the cloud is moving toward the radar site; if the frequency falls, that section of cloud is moving away from the site.

This tool gave researchers their first glimpse at the large-scale motion that takes place in the vast, violent thunderstorms whose broad central cloud columns are dominated by winds rotating counterclockwise. This circulation, which meteorologists call a mid-level mesocyclone, begins about a mile above the ground and can reach diameters of up to 12 miles. Such "supercell" thunderstorms have spawned some of the most violent tornadoes. Researchers suggested that perhaps over time the mesocyclone drops, and when it hits the ground, tornadoes can form.

Questions About Rotation

Yet that left scientists puzzling over why supercells rotate. One suspect was Earth's rotation, which puts a large-scale spin on storms ranging from hurricanes to nor'easters.

In 1978, researchers from the National Center for Atmospheric Research in Boulder, Colorado, and the University of Illinois did what only computer modelers can do: They turned off Earth's rotation. In their simulations, the rotating supercells still formed. Further research showed how low-level winds that changed direction or rapidly changed speed depending on their height generated horizontal whorls of air. These horizontal vortices were drawn into a supercell's central cloud column by strong updrafts, giving the storm its rotation.

This still left scientists up in the air about how rotating air develops near the ground. More field observations and computer time led to the discovery that as rain-cooled downdrafts to the north of the storm sink, they wrap around the warm central updraft, forming a vortex of air along the boundary between them.

By the time these vortices reach the ground, they have evolved into low-level mesocyclones roughly four miles across that tend to get drawn back up into the storm on its southwest side—the region of the storm where tornadoes often form.

"What we don't understand," Dr. Wakimoto says, "is how you get from five or six kilometers to a tornado."

Nor has VORTEX cleared the air.

"Our models don't suggest the kind of [tornado] evolution we saw in VORTEX," says Louis Wicker, a meteorologist at Texas A&M University. For example, he notes that the June 2, 1995, Dimmitt tornado—"a violent, long-lived tornado"—grew amid conditions thought to be twister-killers. "That's a significant wrinkle. It tells us there's something about formation we're overlooking."

New Insights

Doswell notes that a 1994 VORTEX observation tracked an intense tornado that evolved with a fledgling updraft that occurred well away from a preexisting mesocyclone. In that case, a swirling vortex of air near the ground likely was "stretched" by the updraft. As it stretched upward, the vortex narrowed and gained speed, much as a twirling ice skater speeds when she draws her arms close to her body. Indeed, tornadoes are now thought to rise from the ground to the cloudbase; the funnel "cloud" forms only after the tightly whirling column of wind has established itself.

The 1995 observations, Doswell adds, suggest that immediate "triggers" to tornado formation may lie in subtle low-level conditions in and around the storm and may be very difficult to measure. Unfortunately, he says, unless the storm is within 20 miles or so, Doppler radar won't pick up the low-altitude activity.

Meanwhile, as the National Weather Service deploys more Doppler radar, researchers are discovering that more thunderstorms contain mid-level mesocyclones. But they haven't seen as many tornadoes as the number of storms with mesocyclones would suggest.

Where once such storms were thought to generate at least half of all tornadoes, some researchers say this figure could drop to

20 percent or less. Thus, one of the key tornado precursors that Doppler radar was designed to detect seems to be linked to fewer and fewer twisters.

"The challenge is that you're looking for very subtle effects in a storm with 100-mile-an-hour downdrafts, 50-million-volt lightning bolts, and baseball-sized hail," says Tom Grazulis, founder of the Vermont-based Tornado Project, a clearinghouse for tornado information.

Yet for all the uncertainties, he adds, "This is the most exciting time I can remember" in tornado research. Digging into the VORTEX data may well sustain that attitude for years to come.

Storm Chasers

BY JOE NICK PATOSKI

"Storm chasers" are people who follow storms in the hopes of witnessing, filming, and measuring tornadoes. Some are scientists in the field with computers and technical equipment for conducting research and collecting data. Others are partly interested in the science and partly thrilled with the idea of being in the presence of a tornado. Others chase storms purely for the excitement and challenge of it. In this article, Joe Nick Patoski describes his experience of joining the team of Alan Moller, the lead forecaster for the National Weather Service, and Chuck Doswell, a researcher for the National Severe Storms Laboratory in Norman, Oklahoma, and a caravan of meteorologists, students, and "weather nuts" on a chase to spot tornadoes.

Patoski, a senior editor for Austin's Texas Monthly, *reports that the work of storm chasers has improved meteorologists' ability to predict tornadoes and thereby prevent fatalities. Hundreds of people were killed by tornadoes in the 1950s, but the numbers are currently in the tens per year.*

O n the Friday of Memorial Day weekend, I stood on the outskirts of Pampa, high in the Texas Panhandle, feeling like I was in the middle of a battlefield. I was with Martin Lisius and several other storm chasers, the kind of people whose lives inspired the characters in the movie *Twister*. But on this late afternoon, seeing the storm they had tracked down was far more frightening than watching monster twisters surrounded by Dolby sound. All hell was breaking loose as an ominous black cloud hovered overhead, throwing off bolts of lightning that sparked grass fires, which sent billows of smoke hundreds of feet high. We had parked on the road's shoulder, and the chasers were setting up their camera equipment to capture the storm's onslaught of hail, intense downpours, gusts of violent winds, and window-rattling thunderclaps. Close by, I spotted black swirls rising above one of the fires. "Tornado!" I thought and held my

Joe Nick Patoski, "Riders on the Storm," *Texas Monthly*, vol. 24, July 1996, p. 72.

breath. But no, I was told, it was a gustnado, created by a down-draft from within the storm, and it dissipated as quickly as it had appeared.

Amid the fire and rain, the chasers kept their eyes fixed on a wispy dark cloud that emerged from the base of the storm. They were waiting for it to rotate, confirmation that the storm was a mesocyclone, or supercell. As I nervously surveyed the roiling mass, I was again certain a tornado would soon drop. But the storm chasers saw things in the explosive weather that I could not. The edges of the cloud base were not sharply defined, indi-cating that the storm was weakening. When cold blasts of wind shot out from the storm cloud, Lisius, a storm video producer and the founder of the Texas Severe Storms Association (TESSA), jumped back into his Jeep. There wasn't the right mix of warm and cold air needed to make a tornado. As the grass fires glowed in the dusk beneath a flashing, angry sky, we futilely chased clouds south and west for another hundred miles before calling it a day.

That night, at the Travelodge East in Amarillo, a group of chasers gathered in the coffee shop, making plans for the next chase. Around midnight, the waitress asked them if they needed anything else. "Yeah," somebody cracked. "An upper-level dis-turbance."

Living and Breathing Tornadoes

For the past month I had been living and breathing tornadoes, tuning in the weather radio, watching the Weather Channel for hours, and chasing phantom supercells. One day I'd driven from Arlington to north central Kansas in pursuit of a big storm but had seen only blue skies. Conditions, however, were improving. At noon the day after my Pampa experience, about twenty people were gathered in a conference room of the National Weather Service office in Amarillo, studying meteorological data.

"What do you think? Is this a good day for tornadoes?" asked Alan Moller, a lead forecaster for the weather service's Fort Worth office. Many of the mostly male group nodded their heads in agreement. "It's not a sure thing, but it's never a sure thing," an-swered Chuck Doswell, a researcher for the National Severe Storms Laboratory in Norman, Oklahoma. With Doswell and Moller were three of Moller's colleagues, meteorology students from Texas A&M and Michigan Technological University, a cli-matologist from California, and a few meteorologists from the

Weather Channel in Atlanta. The rest were plain old weather nuts. Moller and Doswell obviously had rank. When they talked, everyone listened.

The 46-year-old Moller and 50-year-old Doswell were chasing storms long before Doppler radar, computer modeling, cellular phones, laptops, and the Weather Channel became tools of the trade. They met in 1972, when they were meteorology students at the University of Oklahoma, and have been storm chasing ever since. They schedule their vacations for late May and early June, the best time for tornado outbreaks on the Great Plains, which is the site of more violent weather than anywhere else on the continent.

Tornado Alley

But this spring [of 1996], Tornado Alley—roughly extending from Texas to Nebraska—had been quiet. Tornado Alley is the ideal point of convergence—it is where cool winds in the upper atmosphere collide with warm, moist air from the Gulf of Mexico to cook up the storms that spin off tornadoes. Yet the big storms, born from supercells whose cloud tops reach 60,000 feet, were hammering the Ohio Valley, not Tornado Alley, where the flat, treeless, lightly populated landscape and gridlike roads provide excellent conditions for storm tracking.

On this day the prognosis was good for the area around Amarillo: The polar jet stream had dipped down to provide the shot of cool upper-level winds crucial to break up the dome of warm air, also known as the capping inversion, which hovered at five thousand feet and prevented thunderstorms from forming. Also, a cool front from the north had stalled near the Texas-Oklahoma border and a low-pressure system was moving up from central New Mexico. The Lubbock office of the National Weather Service called it "the most favorable severe-weather scenario for West Texas so far this spring"—good news, if you happen to be a storm chaser. Moller and Doswell and some of the others determined that they had to start in Clovis, New Mexico, to get a piece of the action.

The hour-and-a-half drive southwest on U.S. 60 was uneventful. As the cloud cover broke into scattered "cues," as cumulus clouds are called, Moller's tape deck blasted weather-related blues—"That Mean Ol' Twister," by Lightnin' Hopkins; "Texas Tornado," by Tracy Lawrence; and "Lightnin'," by Johnny Winter. Near Hereford, two dark-colored Ford Explorers with satellite dishes, whip antennas, anemometers on the roof (to measure wind speed), and signs identifying the vehicles as Severe Storm Spotters zipped past doing 90 miles per hour. The lead vehicle, according to the banner on the side, belonged to Warren Faidley, an adviser to *Twister* and the jut-jawed self-proclaimed "World's Only Full-time Professional Storm Chaser." "I used to think Faidley was an okay guy until he started taking himself so seriously," Doswell said matter-of-factly.

The agreeably grumpy Doswell laid out his ground rules: "Don't ask me how many tornadoes I've seen, don't ask me how close I've been to a tornado, and don't ask me if I get scared."

"That's because you're too chicken to get close to one," Moller said, laughing.

Benefits from Storm Chasers

Doswell preferred to talk about the four decades of benefits brought by the work of storm chasers, the National Severe Storms Laboratory, and the National Weather Service despite constant budget cuts from Congress. "In the fifties, tornadoes killed hundreds of people every year. Now it's a few tens a year. But the weather service is suffering from its success. We've done such a good job, no one thinks tornadoes are a threat. To get their attention, tornadoes have to kill people. The history of funding for tornado research follows the history of deaths from tornado disasters."

Although Doswell and Moller rack up more than 10,000 miles on their chasing vacations each spring—shooting storm photographs and video—they're lucky if they spot one or two big tubes a season. For this trip, Moller had added a two-way radio that he bought from Sam Barricklow, a fellow chaser from Dallas and a member of the Dallas-area Skywarn radio storm spotter network. Barricklow was following us, as were several others. But the conversation coming over the radio was irritating Doswell. "Why do I have a bad feeling about today?" Doswell said. "Because everyone's talking about everything but the weather." Just then he was interrupted by a report from the National Weather Service in Amarillo. The entire Panhandle and South Plains were under a tornado watch.

At a convenience store in Clovis the woman behind the counter couldn't help but notice the T-shirts and gimme caps promoting *Storm Track* magazine, *Twister,* and TESSA on the guys in shorts milling around the store and standing out in the parking lot, heads tilted toward the sky. "Y'all aren't tornado chasers, are you?" she asked. She'd just seen *Twister* with her husband. "Don't believe the movie," a chaser told her.

Intercepting the Supercell

In the parking lot, Moller and Doswell identified three cloud structures with supercell potential, watching the cloud towers build, collapse, and reform. "They're all explosive," Moller said, pointing out the well-defined sides of the towers and the glaciation, or icing, of the cloud tops. The storm cloud forming to the north would be hard to track because its projected path was over ranchland with few roads. The two to the south were toss-ups. On a hunch, the chasers decided on the one blossoming closest to us. "These are starting a little early," Doswell said. "Sometimes you can go after one and stay with it. Other times, they're just sucker storms." The storm clouds were moving fast. Moller hollered for the chasers to get into their vehicles and head east in hopes of intercepting the potential supercell along U.S. 84.

"It's prime West Texas tornado country," Moller whooped as we recrossed the state line. "The edge of that anvil is already straight-edged and hard. That sonofabitch is gonna be a supercell!" Four miles later, he suddenly veered off the road to a spot near a tractor tilling the bare red dirt. Five vehicles followed. Five more cars and trucks parked about a quarter mile farther up the highway. The chasers began unpacking cameras and setting up tripods. The storm cloud was a few miles away, and it was producing lightning, thunder, and downpours. Doswell observed a cut in the base, a sign of downdraft that often leads to mesocyclone rotation, but Moller noticed a stratoform, or scud cloud, pushing out from under the base. "That means there is outflow, which is not good for tornadoes," Moller told me. From where we stood, humid wind continued blowing toward the storm, which sucked the energy up. The storm was strengthening. "This is just gorgeous," Doswell said, awed by the spectacle. But the base continued expanding, then contracting. "I'm thinking we ought to go south," Doswell said, but no sooner had we gotten into the car than Sam Barricklow, who was watching the storm from his van nearby, came on the radio. "This storm is starting to intensify and there is slight rotation."

Martin Lisius broke in on the radio. He was north of us and was observing the same storm. He reported that a wall had dropped from the base. Moller pushed the speed limit on the two-lane blacktop, trying to catch up with the storm. Moller saw a banded collar above the base, like a barber's pole, as the wall cloud began rotating, meaning a mesocyclone had formed. The storm was a supercell. "Chuck," he said, his voice rising, "call the National Weather Service. They may have to upgrade this from a watch to a warning from what we can see."

A funnel tail briefly dropped from the low cloud that was rotating violently now. Moller stopped the car. Then the tail retreated, and we resumed toward Bovina, speeding past fields just drenched by a downpour and past toppled farm machinery, indicating wind damage. The radio reported that a tornado warning had been issued. Our car now led a trail of headlights up the road in a storm-induced darkness. Moller madly steered the car through a thick fog rising from just-fallen hail, using his two-way radio to warn others behind us of the road hazard. East and north of Bovina, the wall cloud finally gave the chasers what they wanted, as an imperfect tornado dropped from the mesocyclone's menacing tentacles onto the ground. But no sooner had it touched down than it began to dissipate. "It's getting eaten up by the rear flank downdraft," Doswell said.

Danger Is Clear and Imminent

Outside Friona, we whipped past two vehicles with flashing lights parked on the shoulder—storm spotters monitoring the situation. Danger was clear and imminent. Violent gusts shook the car and knocked off the radio antenna. In Friona the scene was unnerving, as people dashed from their homes to storm cellars. Emergency vehicles were stationed at the town's main intersection. We pulled over at the north edge of town, and Moller and Doswell scrambled for their cameras. Another funnel was dropping from the rotating wall cloud.

"Yes!" Doswell cried as the tornado warning siren in Friona began to wail. "It's well developed . . ." And then a minute later, "Damn! It broke up."

"Let's go get it," Moller yelled, jumping back into the car. "It's gonna do it again."

Sam broke in on the radio. "I see some teeth around the old meso. It may be intensifying." With Moller and Doswell in the lead, the caravan of chasers followed the storm north, then west. Suddenly Moller hit the brakes. "There's a tornado on the ground to the left," he shouted. The cameras focused on the funnel, which was indeed a tornado, though it was practically transparent. Except for the sweet song of a bird on a phone wire, there was an eerie silence. Then Moller yelled again: "We need to back up. It's coming right at us." The translucent cone was only a few hundred yards away and headed in our direction. Moller shifted into reverse, stopping at a safe distance where other chaser vehicles were pulling up.

After about a minute, the funnel fell apart. Up the road, wisps of a low black cloud circulated on the ground near a farmhouse. "That was your quintessential low-level mesocyclone," Doswell said, getting back into the car as other vehicles were still arriving.

"Let's try to think ahead of everyone else," Moller said, speaking into the radio. "Sam, you made a great call on this storm early on, when I thought it was a dead duck." We raced north into Deaf Smith County, passing a truck with flashing lights and a huge radar on the trailer. The National Severe Storms Lab's portable Doppler radar was taking the pulse of the supercell.

A new mesocyclone was developing out of the cloud base. Trying to get ahead of the storm, we drove into Hereford, then zigzagged north and west about 25 miles to intercept the storm. It was two miles ahead of us, but we could clearly make out the funnel's silver-gray elephant's trunk dancing across the range, flashing its power for several minutes before retreating skyward. Watching it filled me with awe and, bizarrely, a desire to get closer. Though nowhere near an F5, which is the most dangerous twister, it was a lethal beauty.

The End of the Chase

We continued our chase for another forty miles until Moller and Doswell reluctantly agreed it was time to give up and start heading south in search of another storm. "Well, we started out this year getting a tornado," Doswell said. "Now we've got to get a photogenic tornado."

"This is where the reality is far better than the fantasy," Moller said.

"You were pretty lucky," Doswell told me, "seeing all that on your first day with us."

At dusk we watched one last storm near Littlefield, vainly trying to make out the silhouette of a rotating meso. Then we headed to the twinkling lights of Lubbock for the night. At the entrance of the Hub City Brewery, Moller and Doswell ran into several other chasers, including Betsy Abrams and Matt Crowther, husband-and-wife meteorologists on holiday from the Weather Channel in Atlanta. They too had started the day in Amarillo, but they had headed down south and west of Lubbock instead of to Clovis. The inevitable question was quickly asked: "See anything?" Abrams reported that they had spotted a gustnado, which they videotaped, and had run into baseball-size hail. That was it.

In the middle of the late-night meal, a wild thunderstorm passed over the city. Moller couldn't restrain himself from going out to watch the action. As if on cue, a lightning bolt hit a pole across the street, briefly turning it a glowing red. A waitress squealed with fright. "Sorry," Moller said. "We probably brought that with us."

Studying Tornadoes

Lubbock was the perfect place for me to end the chase, because with the storm comes storm debris, the subject of study at the Institute for Disaster Research and the Wind Engineering Research Center on the campus of Texas Tech University. The research centers were created after the 1970 Lubbock Tornado caused 26 deaths and $135 million in damage and wrecked the city's tallest building. "We realized that Mother Nature had given us a $135 million laboratory in our own back yard," said James McDonald, the director of the Institute for Disaster Research.

With funding from the National Science Foundation, and later from the Nuclear Regulatory Commission, which was trying to establish baselines for tornado-proof nuclear reactors, the institute has studied wind, tornado, hail, wind shear, and other severe-weather-related events throughout the region, as well as hurricanes on the Gulf Coast. Texas Tech also has a field lab that features a metal building, loaded with sensors, that circles on a railroad track to test wind pressure and wind stress. There is an indoor wind lab with air compressor cannons that fire fifteen-pound two-by-fours, metal pipes, and chunks of hail at speeds of about 100 miles per hour at various wall materials.

Documenting damage from sixty storm events, the centers' studies have yielded some interesting results. Most wind damage is done by weak tornadoes, with winds from 125 to 150 miles per hour. Opening a window in advance of a tornado is no longer recommended because wind damage is caused by the Bernoulli effect, the tendency of wind to speed up as it moves around and over objects. Using reinforced masonry in buildings is essential to avoiding disasters like the 1987 Saragosa Tornado. "It wasn't wind that killed people. It was the falling concrete blocks," said Kishor Mehta, who oversees the wind lab. Since a tornado-proof home is economically unfeasible, Texas Tech researchers came up with the in-home storm shelter, essentially reinforcing the walls and ceiling on an inside room, for less than $2,500. "The mobile home is still very susceptible to tornado damage," McDonald said. "Every mobile-home community should have enough shelters so that no one has to go more than a hundred and fifty feet to be safe."

Even disaster specialists must sometimes answer the call to chase, as Richard Peterson, a meteorologist who works with Mehta and McDonald, did when a farmer near Matador called Peterson after hail the size of grapefruit fell on his land the night before. "He says he's got a three- by five-inch specimen in his freezer, but I've got to come get it today or he is going to let it melt."

Measuring Tornadoes with TOTO

By Howard B. Bluestein

Professor and researcher Howard B. Bluestein recounts the life history of TOTO (TOtable Tornado Observatory), a tornado-measuring instrument built in 1979 to determine wind, pressure, temperature, and electric fields of tornadoes. The four-hundred-pound, barrel-shaped TOTO was taken on numerous tornado chasing trips to test its effectiveness. Bluestein and fellow researchers had glimmers of hope but were frustrated when TOTO was placed in tornadoes' probable paths only to have the twisters dissipate before reaching the site. In the midst of TOTO's testing, Bluestein inadvertently spotted two small tornado-like funnels and coined the term "landspouts," distinguishing these small formations from real tornadoes associated with large supercell thunderstorms. In addition, Bluestein discovered important information on tornado tracking.

After several years of trying, Bluestein decided it was too difficult to put TOTO in the path of a tornado in a timely manner. Others tried, also unsuccessfully, for several years afterward. In 1986, TOTO was decommissioned and placed in the museum of the National Oceanic and Atmospheric Administration's headquarters in Washington, D.C.

Howard B. Bluestein is professor of meteorology at the University of Oklahoma and a visiting scientist at the National Center for Atmospheric Research in Boulder, Colorado. His research has been published in leading scientific journals and he is the author of a textbook on synoptic-dynamic meteorology.

P robably the earliest *in situ* [in position] measurements of tornadolike vortices were made in waterspouts [tornadoes over the water]. During waterspout season in 1970 atmospheric physicist Chris Church and his crew flew over the Florida Keys towing a trailing-wire probe through waterspouts. Four years later, meteorology graduate student Verne Leverson, meteorologist Pete Sinclair, and Joe Golden, flying in an AT-6 acrobatic aircraft, used a gust probe to make measurements below the cloud base of mature waterspouts. They found that the pressure and temperature near the centers of waterspouts are lower and higher, respectively, than outside the waterspout—and they also found that air rises inside waterspouts and sinks outside. Tornado researchers took note of these observations.

In 1972 Bruce Morgan suggested that an armored vehicle could be used to penetrate a tornado to make measurements. This imaginative idea has never been put into practice, perhaps because it is perceived as crazy—a tank would probably inflict quite a bit of damage itself as it tumbles over the countryside, scurrying to get into a tornado. But at the American Meteorological Society's 1979 conference on severe local storms, Al Bedard, a scientist with the National Oceanic and Atmospheric Administration (NOAA) in Boulder, Colorado, approached me with a novel idea. He and his coworkers had built hardened sensors designed to withstand local downslope windstorms, in which gusts can exceed 100 mph, and wanted to build a device that could be placed right in a tornado. It would take wind, pressure, temperature, and electric field measurements. Humidity measurements, which are more difficult to make, were ruled out because we could not find a suitable sensor. At the time, I was only vaguely aware of the earlier work on waterspouts.

With limited funds from NOAA and using spare parts, Bedard and his colleague Carl Ramzy built an ingenious four-hundred-pound device. A barrel-like base housed strip-chart recorders for the wind, pressure, and temperature sensors, which were mounted on a boom above. A sensor for measuring corona discharge, a property of the electrical field, was also mounted on the boom. All measurements would be recorded in analog format. In light of our negligible budget, digital recording would have been too expensive. During the summer of 1980, with the aid of liquid stimulation at a cocktail party in Boulder, we named the device TOTO (Totable Tornado Observatory), after the dog that, along with its owner, Dorothy, was swept away in a tornado in *The Wizard of Oz*. TOTO was apparently the inspiration for a similar device named "Dorothy" in the 1996 movie *Twister*.

How TOTO Worked

TOTO, which was mounted in the back of a government pickup truck, could be deployed in thirty seconds or less. In theory, we would get in the direct path of a tornado, roll TOTO down the truck's ramp, switch on the instrument package, and get the hell out of the way. One does not walk into a tornado holding up an anemometer [a gauge for determining the speed of the wind] and hope to survive! The tornado would, we reasoned, pass over TOTO, probably leaving it somewhat battered but (we hoped) still intact. We would then retrieve our mechanical canine and interpret the data traces. If the tornado did not change in intensity or size as it passed by TOTO, we could convert time to space and thereby determine the profile of wind speed and direction, pressure, and temperature across the vortex. We practiced deploying TOTO just east of the foothills of the Rockies in the vicinity of rather weak thundershowers during the summer of 1980, which was infamous for a searing heat wave in the southern plains of the United States.

The storm season of 1980 had not been eventful in Tornado Alley. For much of May, a blocking ridge diverted atmospheric disturbances elsewhere, and wind profiles were not conducive to tornadic-storm formation. It was not until the end of the month that the ridge broke down, briefly, out in west Texas, and a very strong flow aloft appeared. Severe storms began to brew. Erik Rasmussen, a graduate student at Texas Tech University, shot some impressive tornado movies in his own intercept vehicle, near Tulia and Lakeview, Texas. He also noted the cyclical tendency of some tornado episodes. We were nearing the birthplace of the tornadoes he filmed when our chase car, which had developed a hole in the floorboards, treated us to a geyserlike dousing as we drove into a flooded highway. That ended our chase— we had to limp home, missing the show.

TOTO's First Deployment

We first began to use TOTO on tornadic storms in the spring of 1981, one of the most active storm seasons for us ever. Our field experiment was not yet formally funded and our pilot experiment had been carried out on a shoestring budget. (The weather gods apparently took good notice of the lack of a formal, well-funded field program.) The first outbreak was on May 17, in central Oklahoma. Most of the chase vans from the National Severe Storms Laboratory (NSSL) were focused on making electrical measurements. In our zeal to prove TOTO's worth, we headed north in pursuit of an early tornadic storm. But it was too far ahead to catch, so we retraced our steps, toward Oklahoma City, where we heard that another storm was moving through the northern suburbs. But heavy traffic and blinding rain delayed us, and by the time we got to the right spot, the tornado was long gone.

Next came our worst nightmare. Another storm was forming just southwest of Norman and was easily chased by those fortunate souls who had not jumped out early. While our radio blared graphic descriptions of the tornado, all we could do was listen. But another potentially tornadic storm was taking shape not far away from where we were, and we scrambled toward it. This time we made it and sent TOTO out under a wall cloud to make its first measurements. The storm failed to produce a tornado, but we had finally proved to ourselves that TOTO could be used in a storm.

Almost a week later, on May 22, Bob Davies-Jones at NSSL phoned just after lunch with the news that lightning was being detected in storms that had just popped up in western and southwestern Oklahoma, ahead of the dryline [a boundary where adjoining air masses exhibit extremely low moisture content on one side and extremely high moisture content on the other side]. He suggested that our group leave right then for parts west. I needed a good kick out the door, and Bob provided it.

Not far from overcast Norman, which was under a blustery south wind bearing abundant moisture from the Gulf of Mexico, one of our tires went flat, and we had to stop to change it. Perhaps the low pressure in the tire was a hint of what was to come. The weather gods had dropped a hurdle in our path while storms were building up to our west.

We successfully jumped the hurdle [Such hurdles are not uncommon. In May 1997, I was trapped in an elevator while trying to leave my office to go storm chasing. Fortunately I made it out in time to catch a storm.] and continued our caravan—the parent vehicle, a student's car, and TOTO in its pickup truck. At Cordell, about seventy-five miles west of Norman, we stopped to photograph a wall cloud to the northwest, viewed over a field of waving wheat. What followed was one of the most beautiful life histories of a tornado I have ever seen, timed perfectly for our arrival. The truck carrying TOTO raced northward, but not fast enough to place it in the path of the tornado. Instead, in its first deployment with a tornado in sight, TOTO bore the brunt of strong winds along the gust front south of the tornado, but was not damaged and did not tip over.

A Memorable Day

It was a memorable day. We observed a total of nine tornadoes, more than my group has ever seen since. Although some of that number were only dust whirls with no condensation funnels, and

Researchers had hoped that TOTO, pictured here, would help them determine the wind, temperature, pressure, and relative humidity of tornadoes.

some were too far away to attempt TOTO measurements, one was a monster. The last one of the day and the grand finale, this intense tornado moved through the town of Binger, forty miles west of Norman. It began as a multiple-vortex tornado and consolidated into a large cylindrical one. Fortunately, it dissipated as it headed toward the outskirts of Oklahoma City. We tried to deploy TOTO right in the tornado but succeeded only in placing the device near it. . . .

The chase ended at sunset, just west of Norman. Most of it had taken place along a single east-west stretch of highway. There was no long ride home and no early-morning arrival. It was a big success and our failures the previous week were forgotten.

The Cordell-Binger chase illustrates that you had better not arrive too early on the scene of developing convection along the dryline. It usually takes an hour or more for tornadoes to develop in supercells. If you drive right up to the place where a storm is forming, you run the risk of a tornado forming twenty-five to fifty miles off to the north or south in another storm and you might not be able to get to the right spot to intercept it. It is better to remain well ahead of the dryline, allowing time for the initial line of towering cumulus and developing thunderstorms to mature into individual supercells and then to move north or south to intercept the storm most likely to produce a tornado. The risk of staying too far back is that you may not be able to see the storms taking shape, and without good radar and satellite information you may not know where to head.

Spotting Landspouts

The 1981 storm season was important not only because it marked our first real use of TOTO, but also because of a serendipitous observation. On April 19 we set out, reluctantly, with a photographer and writer from *People* magazine who were eager to get a story about storm chasers. We went out mainly for the sake of the photographer and writer—we knew that the atmospheric conditions in nearby hunting grounds would probably not support the formation of tornadic supercells. But we also knew that in northwest Texas, conditions were more favorable. A storm with a mesocyclone was brewing there, so off we went—even though I felt that the odds of tracking down a tornado in the Texas storm were long. About midway to our destination I turned around to speak with one of our crew in the backseat. Astounded, I spotted a tornado out the *rear* window.

When I called Don Burgess, the operator of NSSL's Doppler radar, from our car phone, he informed me that where we were witnessing a tornado, there was no mesocyclone and hardly any storm! Instead, the radar showed what appeared to be only the beginning of a storm, and not much precipitation was reaching the ground.

Then yet another tornado formed behind us and, like the first, it hovered over open country. Both had very narrow condensation funnels, reminding me of waterspouts I had seen in south Florida years earlier. The tornadoes looked fairly benign; at worst, I thought, they might unroof a doghouse. Subsequently I named small tornadoes in incipient thunderstorms *landspouts* because they look like waterspouts and apparently form under similar conditions. (Later I learned that someone had suggested the term "land waterspout" as early as 1927, but it hadn't caught on.) Landspouts contrast sharply with tornadoes in mature supercell storms, which are usually associated with mesocyclones. I was aware that Don Burgess, in correlating tornadoes with mesocyclone signatures in Oklahoma, recognized that some tornadoes have no such associated signature. Now I had seen two of them myself. The guests from *People* magazine were disappointed that they didn't see a ripsnorting monster tornado flinging debris all over the place. At one point I was asked to pose, point to the tornado, and act frantic: I wasn't, and I didn't.

Landspouts are not the only nonsupercell tornadoes that have been documented. Gustnadoes, [dust whirls], are another. Funnel clouds are sometimes produced in convective storms that form beneath pools of relatively cold air aloft. These are called "cold-air funnels." Nonsupercell tornadoes have also been documented at the leading edge of a rainband along a cold front near Sacramento, California, of all places—California is not well known for its tornadoes.

For the 1982 storm season we obtained funding from the National Science Foundation (NSF) and NSSL to use TOTO in a cooperative effort with NSSL. Our parent vehicle, an old contraption once used by the University of Wyoming, we dubbed the Wyoming van.

On May 11 we initially headed west from Oklahoma City. But after hearing from Conrad Ziegler, the nowcaster at NSSL, that there was a tornadic storm in southwest Oklahoma near Altus, we took off in that direction; unfortunately no well-defined storm structure was to be seen. Soon it became clear that we would have to core-punch the storm—that is, drive blindly through heavy rain and hail to reach the area where there might be a tornado. It was risky, but to take a safer route might mean missing a tornado.

A Golden Opportunity

We now had the precise location and movement of the tornado itself and chose a route by which we could "thread the needle" through the storm, maximizing our chances of intercepting the tornado. If we took a safer route, the tornado might be too far off to see. The rain got heavier and heavier and small hail began to mix in with it. Then the hailstones got larger and the rain disappeared. Finally, there were only spurts of large hail, and visibility suddenly improved. Off to the southwest, several miles away, we could see a decent-sized tornado causing millions of dollars of damage to Altus Air Force Base. Here was a golden opportunity to drop TOTO directly in the tornado's path.

We did so. However, the weather gods realized what was going on and countered by forcing the tornado to dissipate before it reached TOTO. The tease got worse. Another tornado, forming by a process reminiscent of Erik Rasmussen's cyclical tornadogenesis (birth of a tornado) model, appeared to our west. It quickly built into a spectacular multiple-vortex tornado, causing considerable damage to homes in Friendship, a tiny rural town northeast of Altus. Our cameraman, Bill McCaul, grappled with NSSL's 16-mm movie camera, attempting to take what would have been one of the most spectacular tornado movies of all time. Alas, the camera jammed; the tornado, like Medusa, had turned it to stone.

Not Giving Up

Discouraged but not willing to give up, we headed north, hoping to drop off TOTO in the "tornado kennel." The Altus tornado had been moving northeast, as had the parent storm. So it was reasonable to assume that the Friendship tornado to our west would move that way, too. But it didn't—instead, it moved off to the northwest! We continued our chase for an hour, with the funnel in sight and the tornado oscillating between a single vortex and multiple vortices. The closer we got, the farther it retreated westward. I have never watched a tornado for so long, unable to get in its path.

Frustration continued on the next good chase day. On May 18 the Wyoming van broke down near Matador, Texas. Leaving the van overnight to be repaired, we piled into a private chaser's car for the long drive back to Norman. The next day we were able to retrieve our van, now in working order, and to resume our chase. We now had a choice: go toward the Texas Panhandle, where storms that were too far away to see were firing up, or chase an isolated storm that was clearly visible to our west. We decided to target the nearby storm.

The chaotic nature of storm chasing became apparent: Since we had to return to Matador, we were in a better position to chase the "wrong" storm. Had we not gone there, we would have headed straight toward the Texas Panhandle, where the northern storm produced an unusually wide and intense (and now legendary) tornado near Pampa. Tim Marshall and Jim Leonard, private storm chasers extraordinaire, still regale us with tales of "Pampa day." The Pampa tornado was around a mile across, and appeared from a curtain of rain and forced the chasers into strategic retreat. The isolated storm we went after produced virtually no rainfall, rotated, and did not produce a tornado. It was an LP [low-precipitation] storm. . . .

On the last good chase day of the season, June 10, our target area was western Kansas. But intense activity, we heard, was forming in the Northern Texas Panhandle, and we decided to rendezvous with NSSL crews heading there. Once again, as on May 11, we were forced to core-punch, since the storm was to our south. We caught up with Bob Davies-Jones's chase vehicle near Spearman, a town to which we would return a number of times in the following decade. Bob was able to position himself and his crew just ahead and to the east of the wall cloud while we played catch-up, just behind the wall cloud, but just ahead of large hail—a truly precarious position.

Our objective was now to get ahead of the wall cloud as fast as we could and deploy TOTO. A tornado was forming just to our south. Dust whirls were spinning toward us, on a possible collision course. Judging that we could *not* outpace the speed of the tornado, I ordered the chase van and the TOTO pickup truck to stop and deploy. Even if we were just behind the tornado, data collected there could be useful.

The decision was a good one. The tornado crossed the road just to our east, uprooting trees and snapping utility poles. Strong northerly winds blew down power lines, one of which struck the windshield of our van. We sat stock-still, touching nothing, while the power line sizzled, seared a line in the windshield, and bounced off the van. It was a close call. Residents of a nearby mobile home told us that they sought refuge in their underground storm shelter to escape the tornado and returned to find their home completely destroyed. Insulation and twisted sheet metal were strewn about. And so another storm season ended without a direct hit by a tornado on TOTO, the eager tornado dog.

TOTO Is Decommissioned

The next year, in March 1983, we tested TOTO in a wind tunnel at Texas A&M University. Unless it was anchored down or widened, we discovered, TOTO could tip over at wind speeds as low as 100 mph and might not be able to withstand a direct hit by an intense tornado. Even so, we tried again later that spring to make measurements with TOTO unmodified. But 1983 was not a good year for tornadic storms in the southern plains. Only on one day, May 17, was there a large tornado, and it moved over mostly open country, inflicting little damage. Visibility was poor due to blowing dust, and we couldn't get TOTO anywhere near it.

By this time, press and media accounts of what we were trying to do had captured, apparently, the imagination of the public. One day out in rural north-central Oklahoma, just ahead of a storm, we set up TOTO on the front yard of a rural house. The owners must have recognized us and, guessing what we were doing, ran like hell for cover. Of course, the weather gods stepped in, and no tornado formed. Besides, it seemed as though tornadoes tended to avoid TOTO—perhaps the reaction of people seeing us arrive on their front lawn with TOTO should have been one of relief.

On another occasion, we were the ones who ran for cover. Fred Sanders, my mentor from MIT, was visiting and chasing with our team. Fred, who has sailed many times in the Newport-to-Bermuda yacht race, has a store of yarns involving the perils of big waves in tropical storms, nocturnal waterspouts, and other mayhem of the natural variety. I was eager to show him real severe weather storms such as he had never seen. We met a supercell in northwest Texas, but were badly positioned and were forced to core-punch to reach the wall cloud. We drove blindly through its heavy rain core, which eventually turned to large hail, and were lucky to escape without incident. When we broke out of the hail, there, just ahead of us, was a group of storm chasers, all watching us in disbelief!

After the 1983 season we gave up trying to use TOTO; it was just too difficult to get in the path of a tornado at the right time. NSSL tried the maneuver for a few more years, and on April 30, 1985, Lou Wicker and his crew almost succeeded near Ardmore, Oklahoma. Unfortunately, the tornado was just starting up when measurements were made, and the maximum wind speeds recorded were less than those taken south of the Cordell tornado back in 1981. In late 1986 TOTO was decommissioned, and to-day is a museum piece, on view at National Oceanic and Atmospheric Administration headquarters in Washington, D.C. Other approaches would have to be devised for obtaining direct measurements in tornadoes.

Chapter 2

Disasters and Personal Accounts

The Great Tri-State Tornado

By Wallace Akin

The deadliest tornado on record in the United States started in the hills of southeastern Missouri on March 18, 1925. Before it was finished, the tornado had blasted its way through the states of Missouri, Illinois, and Indiana, killing almost 700 people and injuring over 2,000. An average tornado lasts about ten or fifteen minutes and travels around fifteen miles. The Tri-State tornado was on the ground for three and a half hours and traveled 219 miles. It moved forward at an average of 60 miles per hour, twice the average speed for a tornado. It reached 73 miles per hour before it finally dissipated. In the following selection, Wallace Akin, a professor of geography at Drake University, describes the destruction wrought by the Tri-State tornado. He focuses especially on Murphysboro, Illinois, the worst-hit town in the tornado's path. In addition to a high death toll, the twister also caused fires, destroyed buildings and homes, knocked out water and electricity service, and left people confused and wandering the streets.

O n the afternoon of March 18, 1925, a warm day for mid–March, about sixty-five degrees, threatening clouds began to gather in southeastern Missouri, forming a vast dark, menacing super thunderstorm cell. From this blackness a funnel descended, touching down three miles north of the little Ozark town of Ellington. There it killed a farmer, the first of nearly seven hundred who would perish that day in America's most deadly tornado.

For the next three and a half hours, the tornado followed a remarkably straight northeastern course, never leaving the ground. Sucking up huge quantities of debris—dirt, houses, trees, barns— it ejected them as deadly missiles along its route. It cut a path of destruction one-half to one mile wide across three states, Missouri, Illinois, and Indiana. Before its wrath was spent, it had traveled 219 miles, the longest uninterrupted track on record.

My father had just opened a new automobile dealership in the southern Illinois town of Murphysboro. A former mining and farming community of twelve thousand people, Murphysboro had become a bustling manufacturing and railroad center. On that Wednesday at 2:34 P.M., most men were at their jobs and most women were home. As the blackness approached, bells had just signaled the end of recess, summoning children back into their classrooms.

The Tornado Hits Murphysboro

Striking with demonic fury from the southwest, the monster storm smashed its way through the city, killing 234 people and injuring 623, while laying waste 152 blocks and destroying twelve hundred buildings. Water mains burst, electric wires fell, and fires raged out of control. Tall brick school buildings collapsed on students gathered in the hallways. Twenty-five died. Some children crawled from under the debris and in shock wandered home to find no house and, in some cases, no neighborhood. Years later a friend told me that when she reached home, she found only an open field; in the middle of it was her decapitated grandmother, still sitting in her rocking chair.

Searching through old newspapers, I found a remarkable letter published in the *St. Louis Post-Dispatch* four days after the storm. It was written by May Williams, a religious mission worker from the St. Louis area, who was in Murphysboro assisting at a revival meeting held by the Reverend and Mrs. Parrott. Williams wrote her mother: "We left the Logan Hotel about 2:25 P.M. and a goodly crowd was awaiting us in the Moose Hall. Mrs. Parrott opened the service singing More About Jesus. She had sung the first verse and chorus which we were repeating when it suddenly grew dark and there fell upon us what we thought was hail. Rocks began to break through. We were being showered with glass, stones, trash, bricks, and anything. I saw the concrete wall at the back of the hall collapse and come crumbling in. Then the roof started to give way. From outside as well as from within, we could hear terrible cries, yells, screams, and there was a great popping noise. The wind roared—I cannot describe it—and it tore great handfuls from the roof above us. You could see shapes hurtling over us in the air.

Little remains of this Murphysboro, Illinois, church following the 1925 Tri-State tornado, the deadliest on record in the United States.

"Then the storm passed. We went out into the street. We walked the city for an hour or more, terror-struck by what we saw. People went about almost without clothes, with no shoes on, wrapped in rugs or blankets. It was indescribable, the confusion. We picked our way among tangles of wires, trees, poles, brick and lumber to our rooms."

After nightfall, ". . . everything was on fire, it seemed. There was no light except the flare of flames. There was no water. We were black from head to foot." The skin of both living and dead who were exposed to the force of the wind was black from dirt and sand driven into it.

"The fire came closer, and at last we were driven from the hotel and went over to the depot to wait for a relief train. Every place that stood was turned into a hospital. We visited the high school where the doctors were sewing up wounds, giving emergency treatment, and where other helpers were hauling out the dead. We saw numberless torn and bleeding bodies.

"They were dynamiting the city now in their effort to stop the flames, and the roar of the explosions added to the horror of the fires' glare. Everything was ghastly. We had to pick our way to the station by the light of the flames. Then the relief train came. Dead and injured were put on first. We followed."

One of the injured bound for St. Louis hospitals was my father, unconscious, suffering a massive head wound.

Continuing Its Deadly Course

Continuing its deadly, unvarying course, the tornado killed 69 people in De Soto (33 were schoolchildren) and another 31 in rural areas before reaching the largest city in its path, West Frankfort (population 18,500). There it destroyed one-fifth of the city, killing 148 and seriously injuring 410, a toll second only to that in Murphysboro.

The last Illinois town in its path was Parrish (population 270). Arriving at 3:07 P.M., the tornado destroyed 90 percent of the town, killing 22 and injuring 60. There were many heroes during and after that great catastrophe, but none received more gratitude than the principal of the Parrish School, Delmar Perryman. Worried about the stormy weather, he refused to dismiss the 50 or 60 children at the usual time. His decision saved many lives; the school was one of only three buildings left standing. Shortly before 4:00 P.M. the tornado crossed the Wabash River into Indiana and accelerated to an astounding seventy-three miles per hour. As if guided by some malevolent pilot, for the first time it changed course and headed directly for Princeton (population 9,850), its third-largest and final urban victim. Here it demolished 25 percent of the city, killing 45. Finally, at 4:30 P.M., the tri-state tornado lifted and dissipated, its great reservoir of energy spent.

The humanitarian response to the tragedy was immediate, thanks to railroad crews that relayed the news along their route and by telegraph. Medical teams rushed in from near and far, and neighboring fire departments dispatched equipment. Trains carried the wounded to hospitals as far away as Chicago and returned laden with emergency supplies and relief personnel. Throughout the night the medical teams struggled to operate under battlefield conditions, with candles, kerosene lamps, and lanterns providing the only illumination. Supplies of anesthetics, morphine, and antitetanus serum soon ran out. Fortunately, the feared typhoid epidemic did not develop; warnings had gone out to boil all drinking water. Meanwhile, trainloads of coffins arrived from St. Louis and Chicago, along with flowers to adorn them.

The High Death Toll

Why was the death toll so high? Most significant was the lack of any tornado forecast or warning system. The storm moved so fast that people had little time to seek shelter. Few witnesses reported seeing a funnel; they assumed what was approaching was just a thunderstorm.

An average tornado follows a path a few hundred yards wide , and 16 miles long, causing damage across 3 square miles. The tri-state tornado covered 164 square miles. Its intensity, wide path, rapid movement, and long life suggest that it was located near the center of a deep low-pressure system and beneath the core of a strong polar jet stream; the jet kept the storm going by removing air from the top, making way for air to enter at ground level. Also, because of the jet, it maintained its rapid forward motion.

Reading old newspaper accounts, I was startled to discover my father's name on a *St. Louis Post-Dispatch* death list. He was indeed comatose for many weeks and not expected to survive, but he completely recovered and lived to old age. I was a two-year-old when the colossus struck, and as it lifted our house, I went sailing through the air like Dorothy on her way to Oz. Miraculously, I suffered only a minor wound from a piece of glass between my eyes. Like my father, I survived the great tri-state tornado, and I grew up in that town where forever after events were labeled "before" or "after" the storm.

The 1974 Tornado Super Outbreak

BY KELLI MILLER

In the following selection, Kelli Miller describes the worst tornado outbreak in U.S. history, which occurred on April 3, 1974. In less than twenty-four hours, 148 tornadoes touched down in thirteen states, leaving a path of destruction 2,500 miles long from the border of Ontario, Canada, to the state of Mississippi. The twisters left 315 dead and more than 5,000 injured in what has been called "Terrible Tuesday." On the tornado intensity scale, the tornadoes ranked from the lowest to the highest in strength. One of the strongest tornadoes ever witnessed, with winds between 261 and 318 miles per hour, completely destroyed the town of Xenia, Ohio. From the "Super Outbreak," meteorologists learned about the relationship between thunderstorms and tornadoes and more about the phenomenon of tornado "skipping."

Kelli Miller is a freelance television and Web producer who works for the Weather Channel.

I magine 148 tornadoes swarming across some 2,500 miles of countryside one right after another. Such terror undoubtedly makes for good movie fiction, but this incredible scenario is for real.

On Tuesday, April 2, 1974, a storm system with the potential to produce severe thunderstorms began brewing. Meteorologists with the National Weather Service expected the severe activity to strike somewhere in the middle or lower Mississippi Valley. But the tremendous magnitude and intensity of what was to actually occur could not be anticipated.

Thunderstorms began to intensify in the lower Mississippi Valley during the predawn hours of April 3, 1974, while many slept unaware of the impending terror. Dr. Greg Forbes, Severe

Weather Expert at The Weather Channel, was a graduate student at the University of Chicago when the storms occurred.

"I think every meteorologist wants to witness the atmosphere in action," said Forbes.

"So, we all ran up to the roof and watched the rotating clouds. On one hand, we were hoping a tornado would touch down, but on the other hand we were hoping that it did not because of the destruction it could potentially cause."

As the day progressed, thunderstorms intensified dramatically, and more and more severe thunderstorm watches took effect.

The country's worst tornado outbreak began its deadly rampage just after 1 P.M. CDT in Morris, Illinois. Within the next hour additional twisters were spotted in Indiana, Tennessee, and Georgia.

In less than 24 hours, 148 twisters had charged through 13 states stretching from the border of Ontario, Canada, to Mississippi. The tornadoes left a path of destruction more than 2,500 miles long, leaving 315 dead and more than 5,000 injured.

Six of those tornadoes were among the strongest such storms ever recorded. A half-dozen reached an intensity of F5 on the Fujita scale, meaning estimated winds reached or exceeded 261 mph and perhaps were as high as 318 mph. In all, the outbreak included an astonishing 49 killer tornadoes.

Ground Zero for Disaster

Wednesday, April 3, 1974. As school children in Xenia, Ohio, waited for their ride home and workers watched the clock tick slowly towards quitting time, a monstrous tornado whipped wildly towards their small town. In just minutes, the small peaceful city of Xenia became ground zero for the nation's worst tornado outbreak.

The deadly tornado plowed into the Arrowhead subdivision on the southwestern side of Xenia at 4:35 EDT. The tornado, an F5, was among the strongest ever witnessed, with winds estimated between 261 and 318 mph. It sped furiously across town at a speed of about 52 mph.

Frantic residents scrambled for cover as the twister's shrieking winds slammed the historic Xenia Hotel. The tornado showed no mercy—yanking thick trees from the ground, cars from the streets, and people from their homes. It tossed two tractor-trailers 150 feet into the air and onto the roof of a bowling alley. A wooden utility pole about 20 feet long snapped in half like a twig and soared 160 feet away from its original location.

In about the same amount of time it takes a television station to pause for a commercial break, the tornado tore a deadly swath all the way across Xenia.

"Houses, businesses, and landscape that had been so familiar to me had vanished and were now just piles of rubble, not unlike pictures of bombed-out WWII Europe," said meteorologist Don Halsey, who was on duty at the Vandalia Weather Service Office in Dayton, Ohio when the tornado hit.

A hunch by Halsey's co-worker, meteorologist Chester Rathfon, allowed Vandalia forecasters to issue a tornado warning to key points with 18 minutes lead time. Technology in the 1970s made it very difficult for meteorologists to spot the telltale signs of a tornado, called "hook echoes." But Rathfon decided the smudge he saw on his screen was a possible hook echo. Indeed, he was right.

Halsey immediately composed a tornado warning and disseminated it to the media and emergency management. The time on Halsey's teletype message was 1620EDT [4:20 P.M.].

"Fifteen minutes later, Xenia was decimated," said Halsey.

Halsey's first call after issuing the warning was to his two daughters, who lived in Xenia with their mother. "Although I knew the path missed their house, I wanted to make sure they were there and not on their way home from school," said Halsey.

The twister sped out of Xenia as quickly as it entered. As survivors slowly crawled out to survey the damage, nothing could have prepared them for what they saw. The direct hit on their city damaged 2,000 buildings and destroyed 1,300 others. Thirty-three people were dead and 1,600 had been hurt.

"When you see the toys, the clothes, the mattress in the tree, nothing left of the house but the foundation, then the real impact on human life really hits you," said Dr. Greg Forbes, Severe Weather Expert at The Weather Channel.

If any luck was to be found during this disaster, it may have been that the tornado hit after school hours. Had the tornado occurred while school was in session, the death toll could have been even more devastating. Five of the city's 10 schools were nearly demolished; two others sustained minor damage.

"One high school looked as if a bomb was dropped on it," recalled Xenia resident Don Dunstan. "The buses looked like an angry boy threw his toys."

One of the beams from the gymnasium of West Junior High landed 450 feet away from the school. In addition to Xenia's schools, the twister smashed into Central State University in Wilberforce, Ohio, just a few miles to the northeast.

"The picture that amazed me the most was of an automobile rolled up into a near perfect ball," said Halsey. "A young lady from Wilberforce drove directly into the path."

Halsey, Rathfon, and another meteorologist on duty received commendations from the National Oceanic Atmospheric Administration (NOAA) for their actions that fateful day.

"It is possible we did save some lives," said Halsey. "I only wish we could have saved them all. I have never displayed that award."

Tornado Forecasting

Green blobs on the 1957 model radar scopes offered little help in predicting the unprecedented tornadic activity of 1974. Forecasters had to wait for visual confirmation of a tornado before issuing warnings—leaving little or no time for residents to seek safety.

Today, advanced technology such as Doppler radar, weather satellites and sophisticated computer models allow forecasters to issue better watches and warnings. Doppler radar lets meteorologists see strong thunderstorm winds and identify tornadoes before they touch down, giving those in the path of one a better chance of survival.

"What we saw as a green blob on a World War II–vintage radar scope is now depicted in full color and high resolution detail," said John Forsing, Director of the National Weather Service's eastern region.

The super outbreak of tornadoes and imagery from the event offered meteorologists a unique opportunity to study the relationship between thunderstorms and tornadoes.

"The 1974 Super Outbreak was the largest tornado outbreak on record and there was much that we learned from it, meteorologically," said Stu Ostro, Senior Weather Specialist at The Weather Channel.

For example, the tragedy helped meteorologists better understand why tornadoes could demolish one house while leaving a neighboring home unharmed.

"As the result of the Super Outbreak and the research from it, we were able to dispel the previous myth that tornadoes 'skipped'—one house being spared while another was destroyed," said Dr. Greg Forbes, Severe Weather Expert at The Weather Channel.

Instead, researchers learned that suction vortices, smaller intense funnel clouds embedded inside a tornado, hit one house and missed another. Ted Fujita, for whom the tornado's intensity scale is named, theorized the existence of these smaller funnels as early as 1965. But proof of their existence did not come until the Super Outbreak nine years later. Movies taken during the outbreak clearly documented suction vortices inside several additional tornadoes.

"Fujita and I used these movies to reconstruct the wind fields within tornadoes to document just how fast the winds can be in

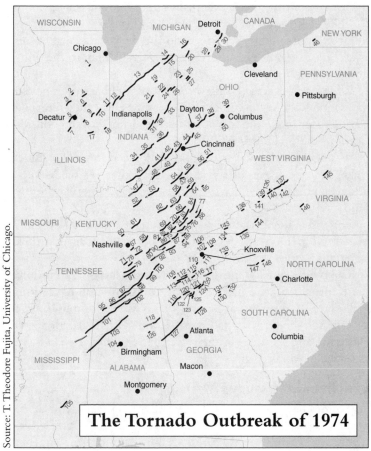

Source: T. Theodore Fujita, University of Chicago.

The Tornado Outbreak of 1974

these killer tornadoes," said Forbes, who studied under Fujita at The University of Chicago.

Forecaster Recalls

When Russell Conger reported for his evening shift as a forecaster intern at the National Weather Service (NWS) office in Louisville, Kentucky, on April 3, 1974, he had no idea he'd soon witness one of the worst tornado outbreaks in U.S. history.

"Severe weather was beginning to break out near Louisville as I reported to work for the 4 to midnight shift," Conger said. "My duty was to record weather observations. Little did I realize I was about to witness one of nature's most spectacular events."

About a half-hour after starting his shift, Conger and his fellow forecasters found themselves uncomfortably close to one of the tornadoes that would eventually snake through the city.

"The electronics technician stuck his head out of the radar room and yelled there was a tornado about two and a half miles away. Ignoring our own safety warnings, we all ran to the windows to look."

When rocks started flying off the roof and hailstones pelted the window, the meteorologists ran for cover. But they returned seconds later to see the tornado over the airport parking lot, adjacent to the NWS office.

Weather records indicate the tornado formed at 4:37 P.M. EDT right "before the eyes of the National Weather Service Meteorologist in Charge," who was giving a live radio interview at the time.

This visual identification allowed Conger and his co-workers to issue a tornado warning 37 minutes before the tornado struck Louisville.

The twister quickly moved away from the airport and toward the Kentucky State Fair and Exposition Center and Freedom Hall on the University of Kentucky-Louisville campus. In just 21 minutes, the tornado had wreaked a trail of destruction 660 feet wide and 22 miles long. More than 900 homes had been damaged beyond repair.

Outbreak Inspires Young Girl

Tornadoes have fascinated The Weather Channel's Kim Perez since she was a little girl. The infamous scene in *The Wizard of Oz*, where a tornado sweeps Dorothy's home up into the air, first spurred Perez's interest in this mysterious weather phenomenon.

This childhood enthrallment instantly became a lifelong passion when a mammoth F5 tornado passed dangerously close to her home on April 3, 1974.

Perez was just 11 years old when the largest tornado event ever witnessed in the United States spawned six massive, F5 tornadoes. One of them slammed into the Sayler Park section of Cincinnati, just 10 or so miles away from Perez's home. She recalls listening to warnings on the radio and tracking the storm on everyday road maps.

"Finally, we watched it from our porch," said Perez. "It was just mesmerizing. My entire family was in awe. But then my dad said 'Get downstairs, now!'"

Perez huddled in the safety of the basement with her family as the monstrosity whisked by, miraculously sparing their neighborhood. But it wasn't until a few days later that Perez fully understood a tornado's incredible fury.

"It was nothing like I've seen before," Perez said as she recalled the aftermath. "To me it was incredible, to see what the wind could do, just devastate and destroy in just seconds."

The experience clinched Perez's desire to pursue a career in meteorology.

"I knew from that point that was what I wanted to do, study tornadoes," Perez recalled. "Ever since that day, I was reading every book I could find on tornadoes."

Perez pored over dozens of library books focusing on severe weather. One book in particular, *Hurricanes and Twisters* captured her fancy.

"I was the only person who ever checked it out," laughed Perez.

Following Her Dream

Years later, still fascinated by severe weather, Perez wrote to a local TV weatherman to learn how he got his start. He told her he joined the Air Force. So that's what she did.

Perez was eventually stationed at Offutt AFB in Omaha, Nebraska, where she worked in the tropical weather section. Although hurricanes and tropical storms did indeed interest her, Perez still aggressively pursued her childhood dream of studying tornadoes.

"There was this other department, called the severe section, but there had never been a woman in that section," Perez said.

But that didn't stop Perez. After persuading her supervisor to give her a shot, Perez became the first woman to serve in Offutt AFB's severe section. She chased tornadoes, spotted a few funnel clouds, but to this day Perez has not seen anything like she saw when she was 11.

Inside a Texas Tornado

BY ROY S. HALL

Roy S. Hall, a retired U.S. Army captain and meteorology enthusiast, tells the story of his family's close encounter with a Texas tornado. After experiencing thunder crashes and tennis ball–sized hail, Hall and his wife realized that the approaching dark cloud, accompanied by a hollow roar, was a tornado. As the monster twister descended upon the family's house, Hall became confused and disoriented. At one point he recognized that he was seeing the inside wall of the tornado where a house wall had been. Looking upward, he could see a thousand feet into the funnel of the twister. Surprisingly, his family survived the disaster relatively unscathed, although fellow townspeople were not as fortunate.

M y wife and I were sitting in our backyard making small talk that warm morning of May 3, 1948, when suddenly she pointed and said, "Look how still those leaves are."

I was startled. The wind was blowing from the south at about 25 m.p.h., but when I looked up at the big hackberry tree I saw what she meant. The wind was so steady and dead-level in its pressure that leaves and small branches were held almost motionless, with scarcely a tremor.

"I'm going in," my wife said. "That solid pressure scares me."

After a bit I went in to have a short rest on my cot. I was barely stretched out when a hard clap of thunder brought my feet to the floor with a slam. The ominous silence that follows close thunder got on my nerves, and I walked through the house to the west lawn to have a look at the weather.

Since noon, thunderstorms had been developing to the west and southwest, muttering and grumbling, miles away, but as the

Roy S. Hall, "Inside a Texas Tornado," *Weatherwise*, vol. 51, January/February 1998, p.16. Copyright © 1998 by *Weatherwise*. Reproduced by permission.

three small clouds that showed prospects of rain were 15 miles off, and drifting north on the air current, I had given them no more thought. The temperature was in the middle-80s, and the air was very humid.

When I stepped off the front porch one of those thunderheads now spanned the western sky, black as ink, less than three miles away. And right across its nearer rim, very low, a mile-long scud cloud [very low, dark, patchy, swiftly moving cloud] was sliding along. The whole cloud was moving swiftly eastward, and had done something I had never heard of before. It had made a right-angle turn in the sky and was cutting across the southerly wind current, which definitely had not slackened. I went to the porch and yelled for my wife.

"What a Terrible Cloud!"

I did not know she had come out till she spoke and scared me. "You sounded urgent, so I hurried the children out . . . Oh!" She had seen the storm. "What a terrible cloud!" I looked around and saw our four children standing on the porch. She said nothing more, but I felt her hand touch my arm in a muted question.

The squall, now about two miles away, was coming directly at us, and the scud cloud, stretched across its front between 400 and 500 feet above the earth, was revolving. Behind the scud cloud a curtain of dark, green rain was falling in a solid, opaque wall.

The south wind was veering. In a matter of a few seconds it was blowing, undiminished, from the southeast toward the cloud. Lightning, the most fearful I have ever seen, and wide as a house, flashed with some regularity between the scud cloud and the ground.

In the comparative stillness following the terrific thunder crashes, I could hear a sustained hollow roaring, like a distant freight train. Feeling my wife's eyes on my face, I said, "Sounds like heavy hail." But it wasn't hail. She knew it too. You can't feel the sound of hail vibrating the air against your face. This was a new sound, one we had never heard before.

The low, deadly looking scud cloud was right on us now, and I saw no sign of a tornado funnel this side of the greenish rain. But it was there, and my wife knew it was there. I told her to go in and take the children. We had no storm cellar, but—had we seen a tornado—we could have gotten into the car and run out of its path. Now we had to take a chance on it missing us. It was

and even, that it resembled the interior of a
. The rim had another motion which I was, for
dazzled to grasp. The whole thing was rotating,
m right to left with incredible velocity.

my left elbow, to afford the baby better protec-
up. It is possible that my eyes then beheld some-
ever seen before and lived to tell about. I was
e interior of a great tornado funnel! It extended
a thousand feet. It was swaying gently, and bend-
rd the southeast. At the bottom the funnel was
across. Higher up it was larger, and seemed to be
h a bright cloud, which shimmered like a fluo-
is brilliant cloud was in the middle of the fun-
g the sides, as I recall having seen the walls ex-
utside the cloud.

where I could observe both the front and back
sides, the terrific whirling could be plainly seen.
tion of the huge pipe swayed over, another phe-
place. It looked as if the whole column were
gs or layers, and when a higher ring moved on
east, the ring immediately below slipped over to
t. This rippling motion continued on down to-
ip.

ve-like motion reached the lower tip, the far edge
s forced downward and jerked toward the south-
in passing, touched the roof of my neighbor's
d the building away like a flash of light, the vari-
g off to the left like sparks from an emery wheel.
stant before, had stood a recently constructed
remained one small room with no roof.

o Vanishes

the rim of the funnel passed beyond the wreck
ng, vaporous, pale-blue streamers extended out
ard the southeast from each corner of the re-
They appeared to be about 20 feet long and six
after hanging perfectly stationary for a long mo-
enly gone.

bluish light was now fading, and abruptly was
was again dark as night. With the darkness, my
come back. I could hear the excited voices of

behind the rain, no question; I had seen them that way.

The low cloud soon passed close overhead, and the dusk of early evening enveloped us. I turned to go in, and as I went up the porch steps hailstones the size of tennis balls began falling on the house and in the yard. My heart sank, for hailstones almost invariably fall in the forefront of a tornado. They came down sparsely, one on about each square yard, but they made a hideous bang and clatter, and I knew some of them were going all the way through our shingled roof. We all went into the west bedroom.

Lightning was striking all around the house now, adding its horror to the fast-rising din. As my wife snapped on the overhead light, a gust of wind and rain hit the room with a crash. My wife was pointing to the west wall. "The wall's blown in!" She had to scream to make herself heard. I could see that the wall had slipped inward six inches or more at the ceiling, and was vibrating under the wind pressure. Drops of water were hitting my face across the room. I tried to assure her. "That gust always comes ahead of a rain squall," I shouted.

The deafening noise outside was growing in intensity by the second, and I realized a tornado was right on us. I yelled in my wife's ear: "Everybody in the back room! Get under the bed!"

A Bit of Hope

On a foolish impulse I ran to the south window for a last look outside before following the family. As I did so the overhead light went off (3:04 P.M., as shown later by our electric clock). It was dark as midnight between lightning flashes, but by shielding my eyes I could see somewhat. I saw my neighbor's house across the vacant lot was standing, but trees and shrubbery out that way were flattened almost to the ground. I could tell the wind was from due west from the direction the planks, sheet-iron, and other debris took as they flailed over the lot. I gathered a bit of hope.

The wind was from the west! It should have been from the south. While a tornado, as a whole, moves generally eastward, the funnel itself rotates counterclockwise, and the west wind indicated we were in the southern edge of the twister. It apparently was passing just north of us. The vivid lightning and rending crashes were also passing on, and there was now a decided lull in the screeching roar outside.

And then very suddenly, when I was in the middle of the room, there was no noise of any kind. It had ceased exactly as if hands had been placed over my ears, cutting off all sound except the extraordinary hard pulse beats in my ears and head, a sensation I had never experienced before. But I could still feel the house tremble and shake under the impact of the wind. A little confused, I started over to look out the north door, when I saw it was growing lighter in the room.

The light, though, was so unnatural I held the thought for a moment that the house was on fire. The illumination had a peculiar bluish tinge, but I could see plainly. I saw the window curtains pressed flat against the ceiling, and saw loose papers and magazines packed in a big wad over the front door. Others were circling about the room, some on the floor and others off it. I came out of my bewilderment enough to make a break for the back of the house.

Some Jarring Blasts

But I never made it. There was a tremendous jar, the floor slid viciously under my feet, and I was almost thrown down. My hat was yanked off my head, and all around objects flashed upward. I sensed the roof of the house was gone.

As I gained footing another jarring "wham" caught me, and I found myself on my back in the fireplace, and the west wall of the room down on top of me.... The side of the room came in as if driven by one mighty blow of a gigantic sledgehammer. One moment the wall stood. The next it had been demolished. I was standing, then I was down, ten feet away. What happened between, I failed to grasp.

By a quirk of fate I was not seriously injured, and as soon as I had my senses about me I clawed up through the wreckage, and crawled around and through the hole where the east door had been. I could tell by the bluish-white light that the roof and ceiling of this room were gone also. I almost ran over my four-year-old daughter, who was coming to find me. Grabbing her up, I was instantly thrown down on my side by a quick side-shift of the floor. I placed her face down, and leaned over her as a protection against flying debris and falling walls.

I knew the house had been lifted from its foundation, and feared it was being carried through the air. Sitting, facing southward, I saw the wall of the room bulge outward and go down. I saw it go, and felt the shock, but still there was no sound. Somehow, I could not collect my senses enough to crawl to the small,

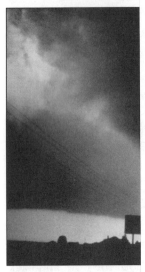

A tornado touches down sou

stout back room, six fe
those pile-driver blasts t

After a moment or so
neighbor's house, standi
yond I could see others
when I saw that we we
been jammed against tre
partly off its foundation

Within the Torn

The relief I experienced
of our house somethin
stood fairly motionless,
It presented a curved f
level. For a second I was
its nature, then it burst
shock: I was looking at
tornado funnel; we wer

The bottom of the r
its leading edge had do
over. The interior of th
pearing to be not over
to the light within the f

side was so sli
glazed standpi
a moment, too
shooting past f

I lay back o
tion, and looke
thing few hav
looking far up
upward for ove
ing slowly tow
about 150 yard
partly filled w
rescent light. T
nel, not touchi
tending on up

Up there to
of the funnel's
As the upper p
nomenon too
composed of r
toward the sou
get back under
ward the lower

When the w
of the funnel v
east. This edge
house and flick
ous parts shoot

Where, an i
home, there no

The Torna

The very insta
of the house, l
and upward to
maining room
inches wide, an
ment, were suc

The peculia
gone. Instantly
hearing began

my family in the small back room, six feet away, and heavy objects falling around the house.

The tornado had passed. The rear edge was doubtless high off the ground, since it went over without doing any damage. Quickly, daylight spread in the wake of the storm, and how good it did look! And how astonishing! In those few long minutes, I had come to believe the tornado had struck at night. It was now only 3:06 P.M.

Luck was certainly with us that day. The only injuries sustained by the family were a severe gash in my boy's arm and a scalp wound on my own head. Others did not fare so well. The tornado cut through the southern part of the city, killing and wounding upward of a hundred people, and causing property damage of over five million dollars.

Trapped in the Great Tupelo Tornado

BY GARY MOORE

Gary Moore was a journalist in Memphis, Tennessee, when his father made a tape recording on which he recounted his experience of the 1936 Tupelo tornado, one of the deadliest in U.S. history. In the following selection, Moore relates his father's story, using quotes from the tape. While driving with his wife, Murray Moore confronted a monster twister face-to-face in the road. Detailing the scenes of destruction that followed, Moore also shares the emotional impact of the event, his panic as the tornado approached, then his loneliness in the eerie silence of the immediate aftermath. He describes his uneasiness as dazed people on the scene jabbered incoherently. He relates the tornado's effects on the rest of the town, where over two hundred people had been killed. Trains made their way through the rubble to carry off the wounded and the dead, leaving survivors with a sense of guilt about not having been able to save them. After such an experience he could not rest easily when storm clouds gathered again and thunder and lightning evoked memories of the chilling disaster.

M y father tried to outrun the Great Tupelo Tornado. The frail Lombardy poplars in the yard were bowing to the wind that Sunday night—April 5, 1936, as people came home from church. Along West Main windows glowed—courtesy of the New Deal. Only a year earlier, President Roosevelt's Tennessee Valley Authority had made Tupelo a feisty little mercantile center in the hardluck hills of Mississippi, the vaunted "First TVA City"—a showcase for big dreams; FDR himself had come to visit in 1935. After April 5, 1936, the dreams would never be the same.

My father peered into the darkened pastures behind the house. He was worried about his mother downtown; he knew the look of tornado weather, and northern Mississippi straddles "Little Tornado Alley," where the twisters grow thick in March and April.

My father was sitting in his new clothtop car in front of the house, racing the motor and waiting for his wife, Ruby. (She was his first wife, not my mother.)

As soon as she closed the car door, they raced east three miles toward town to pick up his mother. The last stretch dropped from rolling limerock into flat open bottomland—where Chickasaw Indians had once planted corn—and here West Main followed a low levee above the bottomland and lay unprotected from the wind. In the dark, they could see nothing.

"The wind preceding the tornado was terribly strong, strong enough to lift the car on one side . . . a frightening thing," my father recalled for my tape recorder nearly half a century later, still awed by his memories.

The car veered crazily across the road, but righted as the gusts ebbed, narrowly missing the abutment of King's Creek bridge. Then they were in town, where lightning strikes illuminated the houses. My father whipped the car onto a side street so he could move north with the wind—fearing they would tip over if he held it broadside.

"The tornado hadn't overtaken me yet, it was still just a strong gusty wind out of the south, but it was picking up things big enough to break the glass out of the windows of the car."

In the Funnel's Path

Then time stopped.

"There was a space of time there that I just couldn't—it seemed like an eternity . . . and then all of a sudden it was like—RRRRROOOOOO!" His hair seemed to stand straight out. The windshield exploded. The car was directly in the funnel's path.

"I knew I was going to die. . . . I thought, If I could just get that door open . . . just get out and run—that's the feeling you got. You just got to get away from it. [The door] had a curved handle and I pushed that thing down, and pushed this way to open it—and it didn't. I put my shoulder against [the door] and pushed as hard as I could." The door didn't budge—probably saving their lives.

"When it was over there wasn't any glass [left]. The car was beat all to pieces." Torrential rain began to fall through the exposed mesh of the roof; lightning flashes revealed broken planks caught in the mesh. An outside door hinge had been driven inward and jutted through the upholstery.

"Get out! Get out!" my father shouted to his wife, not yet knowing the funnel was past—but then he grabbed her arm and pulled her back—he had spied broken and potentially deadly utility wires dangling all around them, torn sinews of the First TVA City.

Lightning continued to slash through the pitch-black night. He looked outside. "Leveled. Flat. Gone." The houses had been replaced by open spaces. There were no human sounds, no moaning, no cries. "It's killed everybody but us," he said numbly.

Fumbling past the debris between them, they asked each other, "Are you all right?" Ruby seemed unhurt. His own hands were cut, his hair gritty with broken glass. They sat trapped among the wires.

An Explosive Emotional Release

"Oh, Lord! Help me, Lord!"

The woman's voice grew louder, a constant, demented wail. Then they heard her footsteps, crashing and stumbling through the wreckage. "Help me! Help me! Help me." Her words came like a rushing train, nearing, then dopplering horribly away. They never saw her.

"She woke everybody up . . . or everybody had maybe grasped a little reality, and people began to moan—and scream and holler in every direction. A flash of lightning came and I saw people congregated over at a little old house."

The house, made of stone, had lost its upper story but was still standing. The sight of living people—after thinking themselves alone in a land of the dead—produced an explosive emotional release.

"I said, 'Open the door! Open the door!' . . . I forgot about the wires! To hell with the wires! Just to get to where there's somebody." The loneliness had been overwhelming. Fortunately, the dangling wires were dead.

The people at the house were buzzing. A woman who had been bathing when the storm hit wore only a sheet. A frail elderly couple had somehow been blown from the second story into the yard—unhurt. Nobody seemed to mind the rain.

But something about the crowd was scary. "They were jab-bering and carrying on, kind of incoherent, talking . . . about crazy things."

"They looked up and said, 'Who are y'all? Where'd you come from?'" They began crowding around the new arrivals and touch-ing them. They would "go over and start feeling you—to see if you're real, you know?" He still shuddered thinking about it.

"I said, 'Let's get away from here. I can't stand this.' So we backed out of there, got back out in the street . . . still trying to get to town. You didn't know which way to go because it was all blocked."

Digging into the Debris

There was a mountain of crushed wood in the street behind the car, apparently the remains of a house. Two little boys sat on top of it. "I can't move," one of the boys said, "my leg is broken." But when they rushed to help him, he said, "Don't help me, help my mother."

"Where is your mother?" my father asked.

The boy pointed beneath him.

"I began to pull a board off, and then before I knew it other people began to come up, and [soon] we were working to-gether. . . .

"We worked down to her. There was his mother and also his father. . . . She was dressed in a gown and he was in his night-clothes, and they were crushed under all that and lying face up on the bed."

They were alive. "She was saying, 'Water! Water!' as though she wanted a drink . . . but it couldn't be that because it was rain-ing as hard as it could rain. But anyway we got her out and put her on a mattress. I suppose there were 10 or 12 people there, helping me, and I'm sure I knew every one of them, [but] I never bothered to look up and see who was helping me."

The woman's husband occasionally gasped for breath. Soon a rescue train would come plowing through the wreckage, pick-ing up the wounded. Unable to get all the way through town, the train would be forced to back its way out, discharging hu-man cargo at whistlestops clear to the hospitals of Memphis, 100 miles away. The parents of the two small boys were on the train; neither survived. Some of the injured taken to other

towns would not be found by grief-stricken relatives for many
weeks.

On Main Street

"We worked our way to a clear street away from the path of the
storm. . . . We went down Main Street. . . . Everybody was
a'milling and a'going." The great water oaks that had graced
downtown Tupelo were gone; the town my generation would
later know would have an austere, unsheltered look.

The high school had been destroyed. The storm had cut a clas-
sic southwest-to-northeast tornadic swath. In the 1960s, my generation would attend junior high in the by-then decaying high
school rebuilt after the storm—a forbidding bunker of squat concrete redoubts, crowned with a tornado warning siren.

My father found his mother safe in the cafe she operated by
the town square—though one of her brothers was briefly feared
lost. He ran a store to the north of the square, on a clay bluff
overlooking Gum Pond, a large backwater. In the storm, many
people were blown from the overcrowded bluff into the waters
of Gum Pond—causing the majority of deaths that night. The
bluffs were a poor area, and many of those who died in Gum
Pond were black.

Tupelo was named for the aquatic tupelo gum trees that grew
at the edges of Gum Pond; now the wreckage of the town lay
scattered on the water's surface.

Tornado Fatalities

The next day, page one of *The New York Times* lamented: "Many
Reported Dead in Tornado; Tupelo, Miss., TVA City, Wrecked."

In his 1953 book, *Tornadoes of the United States*, Snowden Flora
put the number of deaths in the Great Tupelo Tornado at 216.
My father and others in Tupelo recalled the toll being around
236. This was still less than the reported mortality in Little Tornado Alley on May 7, 1840, when a funnel struck Natchez, Mississippi, and reportedly killed 317 people—blowing many of
them into the Mississippi River, an eerie prologue to the Gum
Pond victims of Tupelo.

Even the Natchez storm, however, pales beside the single most
lethal tornado ever recorded in the United States—also in the
Mississippi basin's deadly Little Tornado Alley. For three hours,
the Great Tri-State Tornado of March 18, 1925, cut a swath a

mile or more wide, crossing 219 miles of Missouri, Illinois, and Indiana. It killed 695 people. In the town of Murphysboro, Illinois, alone, 234 died.

In Tupelo, many downtown buildings had been vacant even before the 1936 tornado—thanks to the Great Depression—so there was no shortage of morgue space.

"Just about everybody in town was dead—according to the rumors," my father remembered. "And the next thing you know, you'd walk into people that were supposed to be dead."

A family of 13 my father particularly liked had lived on a limerock knoll behind his pasture. Their house had exploded, and their bodies were strewn across the open land. None survived. My father remembered a wedge-shaped grave, with 13 coffins lining the bottom.

For my father, as for other survivors, guilt was a large part of the wound's immensity. So many people had died. Why hadn't he saved them?

Fear was another part.

"After the tornado, we'd have little clouds come up and it would thunder and lightning and I'd get scared all over again. I would just get desperate to get someplace that was safe." But there was no place to hide.

My father's 120 head of cattle were found among the uprooted trees in the pasture behind the house, most of them dead or dying, the others scoured and bloody. As periodically happens in tornadoes, a barbed wire fence had been uprooted and rolled up, and a two-by-four had been driven through a cast-iron pump cylinder. My father even pulled embedded pine needles out of tree bark, as many dazed survivors elsewhere have done.

One of the survivors of the Tupelo tornado was an infant who lived in a small house on the East Tupelo bluffs, beyond Gum Pond. Born a year earlier amid the electric crackle of the First TVA City celebrations, he had lost a twin brother at birth. In the 1936 tornado, a church near his home was demolished, but the young Elvis Presley survived, though he had heard the howl of the cyclone.

Averting Disaster

Improving Tornado Predictions

BY DAVID WHITMAN, WARREN COHEN,
LAURA TANGLEY, AND CHARLES W. PETIT

Timing is of the utmost importance in predicting potentially deadly tornadoes. Tornadoes usually take an hour to form after the appearance of the giant storm clouds that produce them. However, the weather can change quickly, and threatening weather conditions may arise in half the time meteorologists expect. Scientists have made great improvements in their ability to predict tornadoes and issue warnings. Much of the advancement in tornado prediction is due to a major upgrade in the Doppler radar system, which can actually look inside a thunderstorm and see the tornado forming. In the following selection, the authors illustrate how these improvements helped save lives during a tornado outbreak in Oklahoma and Kansas in May 1999. The National Weather Service (NWS) was able to issue tornado warnings twice as quickly as in previous years. The NWS's goal is to double the average lead (prediction) time for tornado warnings by 2004.

David Whitman is the senior editor for U.S. News & World Report. *Warren Cohen has worked as senior correspondent for Inside.com, a news website. Laura Tangley and Charles W. Petit are writers for U.S. News & World Report.*

Mike Vescio was puzzled. When the meteorologist at the Storm Prediction Center in Norman, Oklahoma, started his 4 P.M. shift on May 3 [1999] his computer monitor showed nothing particularly treacherous. True, just 11 minutes earlier, a severe thunderstorm outlook had been issued. But that was business as usual for Tornado Alley in the spring.

Besides, the powerful jet-stream winds that help spawn dangerous "supercell" thunderstorms and tornadoes were slower than
expected that afternoon.

Fifteen minutes later, Vescio started worrying. New satellite
images showed the growth of several huge clouds, 10 miles in diameter. Doppler radar also detected the presence of mesocyclones
or wind vortexes, a precursor to tornadoes. But for all the technology in one of the country's most sophisticated weather centers, Vescio could only guess whether a tornado was actually
forming. With conditions deteriorating, he issued a tornado
watch. The time was 4:30. The watch area stretched from the
Texas border through most of Oklahoma. That ultimately triggered a series of bulletins on television and radio and an alert to
hospitals to prepare for a disaster. Most tornadoes take an hour to
form after the giant storm clouds appear—if they form at all.
[The May 1999] killer twisters spun up in half that time. "I was
amazed that it happened so quickly," says Vescio. "From that point
on, I knew things were going to get much worse."

Like a Scythe

Vescio didn't have to wait long. Before his shift ended, Oklahoma
and Kansas would suffer one of the nation's deadliest tornado
outbreaks in two decades. More than 70 tornadoes—including a
monster more than a half mile wide—blew through the two
states, killing 46 people, injuring some 900 others, and laying
waste to over 4,500 homes and businesses. Like a giant scythe, the
tornadoes cut down whole neighborhoods, hurling houses and
cars around like chaff. People lost furniture, family keepsakes,
pickups and cars, and scores of pets. But most counted themselves
lucky—they and their loved ones had survived. "It tears you up,"
says Mike Holcomb, who huddled in a closet with his family
while a twister tore off the first and second floors of his house
in Oklahoma City. "But it's not the material things that count."

Inevitably, the deadly twisters triggered questions. Were they
due to global warming, or perhaps La Nina, the vast pool of cool
water in the mid-Pacific that has caused increased precipitation
in the continent's midsection? Could more have been done to
get people to shelter? Do they mark the beginning of a terrible
new round of tornadoes? Most weather analysts say no. [The
May 1999] outbreak seems largely unrelated to long-term cli-
mate shifts, they believe, and doesn't augur an unusually destruc-

tive finish to the tornado season. The giant storm was something of a fluke, in fact—and might have been much deadlier if the early-warning system had not improved in recent years. As far as doing a better job of prediction, the weather watchers say, for the time being they're at the limits of their abilities.

At the heart of the fateful storm was a monster tornado, one that had winds in excess of 260 miles per hour. At times, it was more than a half-mile wide. Storm experts call it an F-5, a rating reserved for "incredible" tornadoes that can strip the bark from trees, lift houses off their foundations, and make automobile-size objects fly more than 100 yards.

Rare Fury

An unholy confluence of meteorological forces gelled to create the tornado. The National Weather Service records about 1,200 tornadoes a year—only one of which, on average, is an F-5. Rarer still is a half-mile-wide F-5 twister that directly hits a heavily populated area. Harold Brooks, a research meteorologist for the National Severe Storms Laboratory in Norman, says "it's been a long time, if ever, since a violent tornado moved through an area with such a large population."

Meteorologists don't know yet why some tornadoes stay on the ground for dozens of miles and achieve the fury of the outbreak in Oklahoma and Kansas while others peter out in just minutes. It is clear, though, that if residents of the Great Plains had not gotten early notice of the tornadoes, hundreds more might have died or been injured. In the Oklahoma City area, television stations started covering the tornado threat more than an hour before the twisters hit, and then announced, almost block-by-block to the tornado-savvy residents, where the twisters were striking.

Mike Holcomb, his wife, 6-year-old daughter, and 9-month-old son cranked up the volume on the TV storm reports before taking their dinner into their walk-in bedroom closet—a typical refuge because few homes in this flood-prone area have storm cellars. When the news reports warned that the tornado was near, Holcomb held a mattress above him as he had been instructed, while his family, covered in blankets, crouched beneath him. "Dear Lord" he began to pray—only to hear a sound like the roar of a train that drowned out his voice. The twister blew away the closet walls and ceiling, spun the family around and sucked off his

The National Weather Service first experimented with the Doppler radar system using this unit.

children's shoes. About all that was left of Holcomb's house was a first floor central closet. Miraculously, no one was seriously hurt.

Throughout Oklahoma and Kansas, hundreds of potential victims survived the storm by taking similar precautions. "We are amazed that we only have five fatalities in our state," said Kansas Gov. Bill Graves. Oklahoma Sen. Don Nickles, who likened the tornadoes' havoc to a B-52 bombing run, said, "If there's only 40 or 50 lives lost in this, that's a major, major miracle." The Storm Prediction Center issued tornado watches 30 to 90 minutes ahead of the storms, and the National Weather Service then put out tornado warnings—showing that a twister had been sighted or indicated by radar—eight to 30 minutes ahead of the first reported touchdowns of the funnel clouds. Eight minutes may not sound like much time, but as late as 1986, the National Weather Service was providing only 4.5 minutes of lead time on average in its tornado warnings. [In 1998], that figure had risen to about 11 minutes. In the race against the clock, extra time saves lives. Television and radio stations can broadcast warnings, officials can sound tornado sirens, and family members can warn friends and relatives.

Much of the improvement in the early-warning system is due to a $1.2 billion weather service upgrade of the Doppler radar system that started in 1992. Pre-Doppler radars could not measure wind fields in storms, and they had less resolution, allowing them to miss some of the small weather fronts where tornadoes can gel. Joseph Schaefer, director of the weather service's Storm Prediction Center, says that with the new network of Doppler radars, "you can actually look inside a thunderstorm and see the bloody tornado forming."

Decades ago, the science of tornado prediction was so poor that forecasters feared they would routinely scare people with false alarms or miss huge tornadoes altogether. Government forecasters were actually barred from using the word "tornado" in weather alerts until 1938 out of concern that they would cause undue public panic. Almost 50 years ago to the day of [May 1999's] storm, a tornado hit Norman, but residents had no warning of the twister until they could see it bearing down on the town. Even a quarter century ago, when 148 twisters killed 330 people in 13 states over a 16-hour period, the weather service still had to wait for visual confirmation of a tornado before issuing a warning.

For all the advances, tornadoes remain one of nature's great mysteries. Just why some severe storms become tornadic while others don't is unclear, and scientists are unable to forecast with any precision the paths that tornadoes will follow. The National Weather Service's goal is to increase the average lead time for tornado warnings to 15 minutes by 2004 by developing more sophisticated software for the Doppler radars and improved computer weather algorithms. Nature's lethal turn through Tornado Alley [in May 1999] was one more sad reminder that every minute counts.

Tornado Safety

BY THE U.S. DEPARTMENT OF COMMERCE, NATIONAL
OCEANIC AND ATMOSPHERIC ADMINISTRATION

*The National Oceanic and Atmospheric Administration (NOAA),
part of the U.S. Department of Commerce, strives to diminish the threat
of tornadoes to life and property. The National Weather Service (NWS),
an element of the NOAA, provides tornado watches and warnings as
well as educating community leaders and the public in disaster preparedness. This selection provides helpful tornado facts such as how fast the average tornado travels, when and where to expect tornadoes, and helpful
advice such as not attempting to flee from a tornado by car. It explains
the important difference between a tornado watch, which indicates where
severe thunderstorms, possibly producing tornadoes, are most likely to occur, and a tornado warning, which indicates that tornadoes have been
spotted in a particular area. Skywarn spotter networks made up of volunteers, police, firefighters, and Civil Defense personnel provide this life-
saving service.*

Their time on Earth is short, and their destructive paths are rather small. Yet, when one of these short-lived, local storms marches through populated areas, it leaves a path of almost total destruction. In seconds, a tornado can reduce a thriving street to rubble.

It is the mission of NOAA, the U.S. Commerce Department's National Oceanic and Atmospheric Administration, to help mitigate the threat to life and property from natural hazards. The National Weather Service, a major element of NOAA, provides the Nation's first line of defense against the awesome destructive force of the tornado. Through its tornado and severe thunderstorm watches and warnings, the National Weather Service gives persons in threatened areas time to find shelter. Further, the National Weather Service, in cooperation with the Federal Emergency Management Agency (FEMA), educates community offi-

U.S. Department of Commerce, National Oceanic and Atmospheric Administration, *Tornado Safety.* Rockville, MD: U.S. Government Printing Office, January 1982.

cials and the public through its disaster preparedness program, on what to do when severe storms threaten.

Tornado Destruction

Every tornado is a potential killer and many are capable of great destruction. Tornadoes can topple buildings, roll mobile homes, uproot trees, hurl people and animals through the air for hundreds of yards, and fill the air with lethal, windborne debris. Sticks, glass, roofing material, lawn furniture all become deadly missiles when driven by a tornado's winds. In 1975, a Mississippi tornado carried a home freezer for more than a mile. Tornadoes do their destructive work through the combined action of their strong rotary winds and the impact of windborne debris. In the most simple case, the force of the tornado's winds push the windward wall of a building inward. The roof is lifted up and the other walls fall outward. Until recently, this damage pattern led to the incorrect belief that the structure had exploded as a result of the atmospheric pressure drop associated with the tornado.

Mobile homes are particularly vulnerable to strong winds and windborne debris. Because they have relatively large surface area to weight ratios, they are easily overturned by high winds. Their thin walls make them extremely vulnerable to windblown debris. Even if tied down, they should be evacuated for more substantial shelter. Mobile home parks should have storm shelters for their residents if located in areas where strong thunderstorms or tornadoes occur.

Tornado Facts

Tornadoes travel at an average speed of 30 miles an hour, but speeds ranging from stationary to 70 miles an hour have been reported. While most tornadoes move from the southwest to the northeast, their direction of travel can be erratic and may change suddenly.

In populated areas, it is very dangerous to attempt to flee to safety in an automobile. Over half of the deaths in the Wichita Falls tornado of 1979 were attributed to people trying to escape in motor vehicles. While chances of avoiding a tornado by driving away in a vehicle may be better in open country, it is still best in most cases to seek or remain in a sturdy shelter such as a house or building. Even a ditch or ravine offers better protection than a vehicle if more substantial shelter is not available.

While hail may or may not precede a tornado, the portion of a thunderstorm adjacent to large hail is often the area where strong to violent tornadoes are most likely to occur.

Once large hail begins to fall, it is best to assume that a tornado may be nearby, and seek appropriate shelter. Once the hail has stopped, remain in a protected area until the thunderstorm has moved away. This will usually be 15 to 30 minutes after the hail ceases.

The tornado's atmospheric pressure drop plays, at most, a minor role in the damage process.

WHAT TO DO IN A TORNADO EMERGENCY

At **home:** Go to the lowest level of the building. If there is no basement, go to an inner hallway or small interior room with no windows, such as a bathroom or closet. Get away from windows. Go to the center of the room; corners attract debris.

In a large building: Go to the basement or an inside hallway on the lowest level. Stay away from auditoriums, cafeterias, large hallways, and other places with wide-span roofs.

Get under a piece of sturdy furniture—a desk, table, workbench—and hold on to it. Put your arms over your head and neck.

If you are in a mobile home, find shelter elsewhere.

Outdoors: Try to get inside. If that isn't possible, lie in a ditch or a low-lying area, or crouch near a large building. Protect your head and neck.

Never try to out-drive a tornado. Get out of the car immediately. Take shelter in a nearby building. If you can't get to a building, get out of the car and lie in a ditch or low-lying area.

U.S. Department of Commerce, National Oceanic and Atmospheric Administration, *Tornado Safety*. Rockville, MD: U.S. Government Printing Office, January 1982.

Most structures have sufficient venting to allow for the sudden drop in atmospheric pressure. Opening a window, once thought to be a way to minimize damage by allowing inside and outside atmospheric pressures to equalize, is not recommended. In fact, if a tornado gets close enough to a structure for the pressure drop to be experienced, the strong tornado winds probably already will have caused the most significant damage. Furthermore, opening the wrong window can actually increase damage.

While most tornado damage is caused by the violent winds, most tornado injuries and deaths result from flying debris.

Small rooms, such as closets or bathrooms, in the center of a home or building offer the greatest protection from flying objects. Such rooms are also less likely to experience roof collapse. Always stay away from windows or exterior doors.

Tornado wind speeds increase with height within the tornado.

Storm cellars or well constructed basements offer the greatest protection from tornadoes. If neither is available, the lowest floor of any substantial structure offers the best alternative. In high-rise buildings, it may not be practical for everyone to reach the lower floors, but the occupants should move as far down as possible and take shelter in interior, small rooms or stairwells.

Tornado winds may produce a loud roar similar to that of a train or airplane.

At night or during heavy rain, the only clue to a tornado's presence may be its roar. Thunderstorms can also produce violent straight-line winds which produce a similar sound. If any unusual roar is heard during threatening weather, it is best to take cover immediately.

Although most tornadoes occur during the mid-afternoon or early evening (3 P.M.–7 P.M.), they can occur at any time, often with little or no warning.

The key to survival is advanced planning. All members of a household should know where the safest areas of the home are. Identify interior bathrooms, closets, halls or basement shelter areas. Be sure every family member knows that they should move to such safe areas at the first signs of danger. There may be only seconds to act. Have a tornado emergency plan at work. Encourage area schools to form a tornado plan and conduct drills.

Tornadoes occur in many parts of the world and in each of the 50 states. However, no area is more favorable to their formation than the continental plains and Gulf Coast of the U.S. during April, May and June. Tornadoes are least frequent in the United States during the winter months, although damaging tornadoes can develop at any time of year.

Watches and Warnings

At NOAA's National Severe Storms Forecast Center (NSSFC) in Kansas City, Mo., National Weather Service meteorologists monitor atmospheric conditions in North America using surface weather observations from hundreds of locations, radar information, satellite photographs, temperature, moisture, and wind speeds in varying levels of the atmosphere, and reports from pilots. Combining these thousands of pieces of information, NSSFC forecasts are able to determine the current state of the atmosphere. When threatening conditions are detected, the work of issuing watches and warnings begins!

A *watch* is issued by the NSSFC to indicate when and where severe thunderstorms and/or tornadoes are *most likely to occur*. A Severe Thunderstorm Watch implies that the storms may develop to significant strength to produce large hail (3/4 inch or greater in diameter) and/or damaging winds. Since all severe thunderstorms are potential tornado producers, a Severe Thunderstorm Watch does not preclude the occurrence of tornadoes. A Tornado Watch means that conditions are favorable for the occurrence of both tornadoes and severe thunderstorms.

Watches are usually issued for areas about 140 miles wide and 200 miles long. During a Tornado Watch, everyone in or near the Watch area should be alert for signs of threatening weather and make preliminary plans for action. Listen to NOAA Weather Radio, commercial radio, or television for further information.

Warnings are issued by local National Weather Service Offices when severe thunderstorms or tornadoes are indicated by radar or reported by trained spotters or other reliable sources. While radar is an invaluable tool, it cannot be relied upon totally to identify all severe weather. Trained spotters with rapid communication capabilities are critical to the warning process. *Skywarn* spotter networks are composed of volunteers such as Amateur Radio and Citizen's Band radio groups as well as emergency service units such as police, firemen and Civil Defense personnel. The *skywarn* spotter networks are the backbone of the warning service and save many lives each year.

Yet, even the best combination of radar and trained spotters cannot identify all tornadoes and severe thunderstorms. If you see a tornado or funnel cloud, report it immediately to a local government agency and ask them to relay your report to the National Weather Service. Tornadoes can occur without warning, and your report and actions could not only save your life but others as well.

A *warning* will describe the area at risk from a tornado or severe thunderstorm. This is determined from the location, size, direction, and speed of movement of the storm (which can be erratic). A severe thunderstorm presents the danger of damaging winds, large hail (three-fourths of an inch in diameter or larger), lightning and heavy rainfall. Listen closely to the information contained in the warning. If a tornado is nearby, take protective cover immediately. At times, you may be in a warning area, but the reported tornado may not be nearby. Remember, you may still be at risk and should be prepared to take cover since the storm may be moving your way or may even produce additional tornadoes or damaging winds.

Tornado Emergency Management in Tornado Alley

BY BETH WADE

In the following excerpt, Beth Wade describes tornado preparedness plans created for the area most hit by tornadoes, "Tornado Alley," which includes Texas, Oklahoma, Kansas, Nebraska, and South Dakota. A generic emergency plan has been developed that can be tailored to cities and towns anywhere. Emergency planners enlist volunteer ham radio operators as spotters in the field to verify the reports of weather radar. Technologies involved in the warning process include radar systems, warning sirens, radio, television, and multiple weather information systems. In addition, emergency managers give safety seminars to schools, businesses, and governments so that the public does not become complacent in a quiet tornado season.

Beth Wade is the managing editor for American City & County.

As threatening weather descended upon Finney County, Kansas, in 1995, County Emergency Management Coordinator Dave Jones dispatched spotters to relay field reports on worsening conditions. "Three spotters were calling in reports about wind velocity and prevailing conditions," Jones says. "One reported 45 mile-per-hour winds from one direction, and his teammate reported the same wind speed from the opposite direction. The one in the middle reported dead calm and said he was looking up into a doughnut."

Whether they come as brief downbursts or as raging clouds

spinning at hundreds of miles per hour, tornadoes are destructive, often sudden and always dangerous. People who have witnessed them know first-hand the eerie calm before the storm; the ascending assault of lightning and hail; and the winds that rip toward earth with the sound of a freight train and the power of nothing else known in nature.

Few know the experience better than the residents of Tornado Alley, the column of states encompassing Texas, Oklahoma, Kansas, Nebraska and South Dakota. There, dry arctic air meets the warm moisture of the Gulf Stream and explodes over the plains with frightening regularity. In Kansas alone, there are 48 tornadoes every year; squeezed into a "tornado season" that lasts from April through September, that number averages out to eight twisters per month.

Surprisingly, of the more than 120 tornado-related deaths in the United States in 1998, only two occurred in Tornado Alley. That may be partially attributed to the region's emergency management efforts, which combine response planning, human resources, technology and public education to enable communities to protect lives that are otherwise helplessly placed in harm's way.

Establishing a Plan

"I'm just amazed that people really don't know what emergency management is," says Jon Tilley, emergency management director for Claremore and Rogers County, Oklahoma. "You say Federal Emergency Management Agency [FEMA] and they say, 'Oh, yeah.' But they don't understand the local part of it—that local efforts make it work." That local effort begins with a uniform response plan, he says.

Prior to 1997, Tilley's response plan, like that of every other community in Oklahoma, was exclusive. "Each community had a plan, and we had enough complaints from city and county directors that the state came out with a plan that is more or less generic," he explains. "Now, every city and county in Oklahoma has it."

Covering operations hierarchy and including law enforcement, fire and rescue, maintenance, communication, health and medical resources, and damage assessment, the plan has to be tailored to some degree (e.g., to include a mayor instead of a commissioner or a maintenance foreman instead of a public works director), but it remains essentially the same across the state. "A lot of times, we assist other counties, and we know their protocol," Tilley says. "We all know how the plan works, so [the uniformity] really does help."

Applied to tornadoes, the plan begins with an eye on the sky. "My operations manager and I both monitor the weather," Tilley explains. "We have two types of radar, or, sometimes [when I'm at home], I'll get a call from the sheriff or the National Weather Service that says the area is under a severe thunderstorm warning."

"As the storm gets closer, we come to the emergency operations center (EOC), and, if it's a tornadic storm, I'll page my staff (including three communications officers and a secretary)," he continues. "As the storm gets more violent, I page the spotters and send them to a spot to wait. At the same time, we get on our commercial or county radio and call the small towns around the area and notify their fire departments. We also notify our electric companies."

If a tornado touches down, Tilley's job shifts into that of a co-ordinator. "You're sort of a Radar O'Reilly," he says. "I'll start notifying outside resources such as the sheriff and medical [personnel] to respond. We make sure that the incident commander gets what he needs and that everything runs smoothly. We make sure there's coordination with all agencies so that the right hand knows what the left hand's doing."

In Sedgwick County, Kansas, the progression is similar. "We've got people in the field, and we're talking to the National Weather Service," says John Coslett, the county's emergency management director. "If a tornado is imminent, we activate our EOC to a Level 2, which brings in all emergency services."

Sedgwick County's resources include city (Wichita) and county law enforcement, fire departments and public works departments; a countywide ambulance service; and the county coroner/medical examiner. Additionally, the EOC provides work stations for the American Red Cross, the Salvation Army and the local school district, which supplies shelter space in the event of a storm.

Coslett notes that local attention to tornado response and co-ordination came after a 1991 tornado that caused 13 deaths in a nearby community. "We had about three different areas where we had losses in our county, but [most of] the losses of life were in the next county," he says. "We were fortunate. It had been years since anything had happened around here. We managed to get through it in a relatively good manner, but we got together with all the emergency services and support people and figured out how we wanted to run this thing in the future."

As a result, Sedgwick County updated its EOC and invested in training with the Emergency Management Institute, Emmetsburg, Iowa. Seventy-eight government and nongovernment personnel attended the three-day workshop in March 1993. "It's site-specific," Coslett explains. "The institute sends somebody into your community, and they spend a week or so driving the community and looking at everything, and they design an exercise based on your community."

For the first two days of the workshop, participants talked about the problems they faced under the existing emergency management program and worked on building solutions. "The big benefit is that you have [everyone involved in the program] sitting in one room and hearing exactly the same thing," Coslett says.

The final day, attendees took part in a mock emergency exercise. "It was a tremendous learning experience," Coslett notes. "You realize that you have to work together when something's going on and that there's plenty of stuff to be done for everybody. You try to get rid of the turf battle type stuff."

Enlisting Volunteers

For almost every emergency management department, the core staff is small—perhaps a director and assistant director—and it is

Keeping the public informed about tornadoes can significantly reduce the number of lives claimed by even the most devastating tornadoes.

augmented by support services and volunteers. Reserve fire-fighters are an essential part of the team, as are amateur radio operators.

In terms of drama, the ham radio operator has one of the most intense roles in tornado response. As a storm approaches, teams of spotters are sent out to provide "ground truth infor-mation" or to confirm what is showing up on weather radars.

"We have two volunteer radio groups that work with us," Coslett says. "So, if the National Weather Service calls and says they need spotters in the field, we notify the heads of the groups, and they send people to the area(s) that the weather service would like them to look at."

Although they work under dangerous conditions, spotters are not storm chasers, Finney County's Jones notes. "The movie *Twister* distorted the facts to the point of being dangerous," he says. "Obviously, we don't want folks trying to chase storms." (Nevertheless, the middle United States has an established indus-try of private storm-chasing tours.)

"The spotters are put in key positions that we look at through radar," Tilley adds. "We might say, 'You're going to be receiving hail. We need to know how big the hail is, and we want to know wind speed or wind gusts.'" That information is then relayed to the EOC via radio.

He is quick to note that weather spotters for Claremore and Rogers County attend an annual training session provided by the National Weather Service. "They are required to go to this school, or they do not spot for us that year," he says.

Incorporating Technology

While the spotter's work hints at the necessity for human and technological interaction during a tornado, preparedness and warning drive the point home. Radar systems, warning sirens, ra-dio and television all are part of the arsenal for saving lives.

"If there's any kind of severe weather in the area, we're mon-itoring it constantly," Coslett says. "We have a direct line to the National Weather Service, and we have live radar from one of the television stations here." Sedgwick County's EOC also receives radar reports from the Emergency Manager's Weather Informa-tion Network and the Data Transmission Network.

Jones and Tilley report similar capabilities in their EOCs. "We've got information from six or seven types of weather information systems before anybody's even out in the field," Tilley notes.

As radar data is confirmed and tornado warnings are issued, residents are notified via local radio and television stations. However, many communities have taken severe weather warning a step further by installing sirens.

"Sirens are made to be outdoor warning signals," Coslett says. "They're not made to be heard in your living room while you're watching television. The main thing is that, when the sirens go off, people should get to a radio or television and find out what's going on and where to go from there."

To ensure that the area's sirens work properly, Coslett's staff tests them every Monday at noon. "After we sound them, we call the residences or businesses located next to the sirens and check whether they are sounding/operating properly," he explains.

Tilley notes that, while sirens can be helpful warning tools, they have limited reach. "Each town is more or less responsible for its own warning system, and that leaves the rural area without it. So we've picked up a repeater and put our emergency management frequency on it, and that allows us to notify people over the radio and through their scanners at home. Also, if an area's got enough time to be warned, we'll have the sheriff's department go through those areas with their sirens."

Preparing the Public

Although emergency managers are charged with the task of doing everything they can to notify residents of impending danger, residents have the responsibility of heeding the warning, Jones says. "People in this part of the country are well-conditioned to tornadoes," he notes. "We ask them to take warnings seriously."

To ensure preparedness for and safety during a tornado, emergency managers have added public education to their list of duties. "We give public education programs on where people should shelter," Jones says, adding that discussions about tornadoes should reach beyond the fear of funnel clouds. "Lightning and floods kill more than tornadoes. Tornadoes tend to receive press."

"We go to schools, businesses, governments and talk to them about safety," Coslett says. "We may look the building over and make recommendations as to what would most likely be the safest part of the building."

"We talk about the difference between a tornado watch (conditions are conducive to the formation of a tornado) and a tornado warning (a tornado has been spotted)," Tilley notes. "We talk about what to do if you have a lot of lightning. Or we even go into nursing homes and tell them about water shortages and storing water." His department also encourages residents to purchase flood and fire insurance.

In addition to safety seminars, public education may encompass written materials and media coverage. Coslett notes that Sedgwick County's brochures are printed in English, Spanish and Vietnamese to accommodate the area's diverse population. Additionally, he notes, local newspapers and television stations pick up educational events throughout the year.

Honing the Response

Even with an established plan, trained volunteers, technology and public education, the tornado response plan is only as good as the person in charge, Tilley notes. "A lot of plans are put up on the shelf and never looked at again," he says. "I could tell you war stories."

To know the plan and ensure its usefulness, the emergency manager must test it on a regular basis. "I do a city drill and a county drill annually," Tilley says. "I call a meeting with [all the emergency support services], and we discuss the parts of the plan that they want to test.

"The police department may say, 'We've got this new system, and we'd like to check it out' or 'I've got a new dispatcher, and I'd like to see how he's going to work under pressure.' I get their input, and then I write the drill.

"We're not out there to say, 'Oh, you did it wrong,'" Tilley adds. "We're out there to learn and find out how we can better [our response]."

The very essence of tornadoes' randomness and destruction dictates that tornado response will never become second nature to emergency managers. However, honing and practicing the plan can bring a sense of predictability to a situation in which chaos is sometimes the only thing remaining intact.

As testimony from Tornado Alley illustrates, a variety of resources can assist in tornado preparedness and reaction. Effectively coordinated, they can reduce losses and save lives when nature's fury is unleashed.

Using Microwaves to Tame Tornadoes?

By Suzanne Mengel

California physicist Bernard Eastlund is developing a plan to prevent the formation of tornadoes. As explained by freelance science writer Suzanne Mengel in the following selection, Eastlund's idea is to blast the potential tornado-producing cloud with about 100 million watts of microwaves from satellites orbiting the earth. The satellites would have solar panels on them that would collect energy from the sun and convert it into microwave beams. In the formation of a tornado, warm ground air rises and spirals then hits a barricading wall of cold air and is forced back downward. The air spiraling back downward forms what is called a mesocyclone. If the mesocyclone touches the ground it may become a tornado. Microwave beams would help warm the cold air and keep it from blocking the rising warm air so that a mesocyclone would not be formed. Eastlund hopes to test his theory sometime in the next decade.

Bernard Eastlund has a score to settle. Back in 1982, Eastlund had just moved to Houston, Texas, when a tornado hit his property. The tornado didn't hurt him or his house, but it mowed down all his pine trees. "That was scary enough," he said.

Eastlund, a physicist and head of the Eastlund Scientific Enterprises Corporation in San Diego, is now plotting his revenge. He says microwave beams shot into thunderstorms can zap tornadoes before they're even born.

Suzanne Mengel, "Taming Twisters," *Current Science*, vol. 86, April 20, 2001, p. 10.

Birth of a Twister

Eastlund believes he can prevent tornadoes by halting the formation of mesocyclones—huge, swirling columns of air that occur during thunderstorms. In a thunderstorm, warm, humid air near the ground rises. As the air moves upward, it starts to spiral. At a certain point high above the ground, the warm air runs into a layer of colder, heavier air. That cold layer acts as a barricade, forcing the spinning air back down again to form a mesocyclone. If the mesocyclone touches the ground, it becomes a more tightly coiled funnel of spinning air, or a tornado, and mayhem may ensue.

Mesocyclones might be stopped, says Eastlund, by busting the cold air barricade with microwave radiation—the same electromagnetic waves you use to nuke popcorn in a microwave oven. According to Eastlund's calculations, raindrops in the cold barricade would absorb the microwaves and release as much as a billion watts of energy. That energy would heat the surrounding cold air and smash the cold barrier the way dynamite demolishes a concrete wall. With no barrier in place, warm, spinning air would continue rising instead of being forced back down to form mesocyclones and tornadoes.

Space Beams

How on Earth could a microwave beam nuke a mass of air? Not from Earth, but from space. Eastlund proposes having Earth-orbiting satellites do the job. Solar panels on the satellites would collect energy from the sun and convert it to microwave beams. Eastlund calls his proposed satellites Thunderstorm Solar Power Satellites (TSPS).

The National Aeronautics and Space Administration (NASA) toyed with a similar idea in the 1960s as a way of creating an alternate energy source for Earth. NASA planned to have microwaves beamed down to receiving stations on Earth, where the beams would be converted to electricity.

Eastlund first came up with his idea for blasting tornadoes during the mid-1980s while working for an oil exploration company in Alaska. At the time, the U.S. government was exploring a plan called the Strategic Defense Initiative (SDI) to shield the country from nuclear attack.

One idea for SDI involved opening a missileproof umbrella of high-energy electrons over the United States. Eastlund suggested erecting large microwave antennas, powered by Alaska's huge natural gas reservoir, that would fire microwaves into the ionosphere. The ionosphere is a layer of the upper atmosphere full of charged particles. The microwaves and the ionosphere's charged particles would interact and release hordes of electrons. Those electrons, attracted by Earth's magnetic field, would form a missileproof dome over the United States.

The SDI plan never went further than the development stage. So Eastlund began pursuing other applications for microwave beams.

Eastlund teamed up with colleagues at the University of Oklahoma's Center for Analysis and Prediction of Storms. There, they created computer simulations of violent weather conditions, then ran the simulations on a Cray C90, the world's fastest type of supercomputer. The simulations helped Eastlund determine how strong a beam he would need to zap a mesocyclone.

Test Run

Sometime in the next decade, Eastlund hopes to test his theory. To do that, he says, he will need a sophisticated Doppler radar system that can look downward from a satellite and locate mesocyclones within thunderclouds. He will also need access to instruments on board the International Space Station to create minibursts of microwaves to test whether they have enough power to heat even the slightest amount of air in a storm cloud.

Some scientists are skeptical of Eastlund's idea. Harold Brooks, a research meteorologist at the National Severe Storms Laboratory in Norman, Okla., doubts tornadoes can—or should—be tamed. "If we take that option away from the thunderstorm, it may respond in some way that is even less friendly than a tornado," he said.

Eastlund is optimistic, though. He hopes that by the time he's ready to start running experiments, a new generation of scientists and engineers will be helping him out. Already, he's heard from one interested middle school student. Eastlund helped the boy create a minimockup of his satellite system. "One of my proudest moments," said Eastlund, "was [when] the 11-year-old [got] an A+ on his science project."

It's payback time for tornado survivor Bernard Eastlund, who wants to stifle tornadoes with microwaves.

The Fujita Tornado Intensity Scale

Scale	Wind Estimate (MPH)	Typical Damage
F0	<73	**Light damage.** Some damage to chimneys; branches broken off trees; shallow-rooted trees pushed over; sign boards damaged.
F1	73–112	**Moderate damage.** Peels surface off roofs; mobile homes pushed off foundations or overturned; moving autos blown off roads.
F2	113–157	**Considerable damage.** Roofs torn off frame houses; mobile homes demolished; boxcars overturned; large trees snapped or uprooted; light-object missiles generated; cars lifted off ground.
F3	158–206	**Severe damage.** Roofs and some walls torn off well-constructed houses; trains overturned; most trees in forest uprooted; heavy cars lifted off the ground and thrown.
F4	207–260	**Devastating damage.** Well-constructed houses leveled; structures with weak foundations blown away some distance; cars thrown and large missiles generated.
F5	261–318	**Incredible damage.** Strong frame houses leveled off foundations and swept away; automobile-sized missiles fly through the air in excess of 100 meters (109 yards); trees debarked; incredible phenomena will occur.

Developed in 1971 by T. Theodore Fujita of the University of Chicago.

The Ten Most Deadly U.S. Tornadoes

Rank	State(s)	Date	Dead	Injured	Town(s)
1	Missouri Illinois Indiana	March 18, 1925	695	2,027	Murphysboro, Gorham, DeSoto
2	Louisiana Mississippi	May 7, 1840	317	109	Natchez
3	Missouri Illinois	May 27, 1896	255	1,000	St. Louis, East St. Louis
4	Mississippi	April 5, 1936	216	700	Tupelo
5	Georgia	April 6, 1936	203	1,600	Gainesville
6	Texas Oklahoma Kansas	April 9, 1947	181	970	Glazier, Higgins, Woodward
7	Louisiana Mississippi	April 24, 1908	143	770	Amite, Pine, Purvis
8	Wisconsin	June 12, 1899	117	200	New Richmond
9	Michigan	June 8, 1953	115	844	Flint
10	Texas	May 11, 1953	114	597	Waco

GLOSSARY

air mass: A large body of air in which the horizontal temperature and moisture distribution is fairly uniform.

air pressure: The weight of the atmosphere.

anticyclone: A high-pressure area in which winds rotate in a clockwise direction in the Northern Hemisphere.

anvil: A cloud of tiny ice crystals that extends around and downwind from the top of a thunderstorm.

condensation: The process by which water vapor is converted to water, releasing heat.

convection: The transfer of heat upward into the atmosphere that frequently creates thunderstorms.

cumulonimbus cloud: A tall cloud associated with thunderstorms and **tornadoes**, sometimes called a thunderhead.

cyclone: A **low-pressure area** in which winds rotate in a counterclockwise direction in the Northern Hemisphere.

dew point: The temperature to which the air must cool in order for **condensation** to occur.

Doppler radar: A type of **radar** that measures both the position and the velocity of objects.

downburst: Potentially damaging winds moving straight down from a thunderstorm and then producing a horizontal zone of influence greater than two miles wide.

hook echo: A **radar** precipitation pattern that looks like a hook or the numeral 6, which sometimes indicates the presence of a **mesocyclone** and **tornado**.

jet stream: The core of high-speed winds that can exist at almost any level of the atmosphere.

low-pressure area: An area where winds flow around a central point in a counterclockwise manner in the Northern Hemisphere and where the net movement of air is upward.

mesocyclone: The spinning, rising air of a thunderhead that can produce a **tornado** under the right conditions.

meteorologist: An individual involved in the study and/or forecasting of the weather; requires a university degree in meteorology or a degree in math or physics with the appropriate meteorology classes.

multiple vortex: The situation in which a **mesocyclone** produces several **tornadoes** at the same time.

overhang: A **radar** feature that helps to designate a thunderstorm as a **supercell**.

precipitation: Liquid or frozen drops of moisture that fall from the atmosphere.

radar: An acronym derived from *r*adio *d*etecting *a*nd *r*anging; technology used in meteorology to detect precipitation intensity and speed of movement.

scud clouds: Low, dark, patchy, swiftly moving clouds.

skipping: The tendency of **tornado** damage to be random as the tornado skips along in its path.

squall line: A narrow line of thunderstorms.

suction vortices: Secondary **vortices** that are theorized to be part of a tornado **vortex** and that may be responsible for the skipping behavior of tornadoes.

supercell: A thunderstorm that exhibits certain well-defined characteristics and usually produces severe weather.

tornado: A powerful, twisting windstorm that begins in the air currents of a thunderhead and touches the ground.

tornado cyclone: A **mesocyclone** inside a thunderstorm that may produce a **tornado**.

tornado warning: A warning issued when a **tornado** has been sighted or is indicated on **radar**, covering a small area and usually in effect for an hour or less.

tornado watch: A warning issued when there is a chance of **tornadoes**, covering hundreds of square miles and usually in effect for several hours.

vortex: A **tornado** or the center of circulation of any **low-pressure area**.

vorticity: The tendency of an air mass to rotate.

wall cloud: The characteristic cloud that forms at the base of a thunderstorm before a funnel cloud appears.

Books

R.F. Abbey Jr. and T.T. Fujita, *The Thunderstorm in Human Affairs*. Norman: University of Oklahoma Press, 1983.

L.J. Battan, *The Nature of Violent Storms*. Garden City, NY: Doubleday, 1961.

Howard B. Bluestein, *Tornado Alley: Monster Storms of the Great Plains*. New York: Oxford University Press, 1999.

K. Davidson, *Twister: The Science of Tornadoes and the Making of an Adventure Movie*. New York: Pocket Books, 1996.

Peter S. Felkor, *The Tri-State Tornado: The Story of America's Greatest Tornado Disaster*. Ames: Iowa State University Press, 1992.

Snowden D. Flora, *Tornadoes of the United States*. Norman: University of Oklahoma Press, 1953.

Thomas P. Grazulis, *The Tornado: Nature's Ultimate Windstorm*. Norman: University of Oklahoma Press, 2001.

Jonathan D. Kahl, *National Audubon Society First Field Guide: Weather*. New York: Scholastic, 1998.

———, *Storm Warning: Tornadoes and Hurricanes*. Minneapolis: Lerner Publications, 1993.

W.S. Lewellen, *The Tornado: Its Structure, Dynamics, and Hazards*. Washington, DC: American Geophysical Union, 1993.

E.N. Lorenz, *The Essence of Chaos*. Seattle: University of Washington Press, 1993.

Michael H. Magil and Barbara G. Levine, *The Amateur Meteorologist*. New York: Franklin Watts, 1993.

J.M. O'Toole, *Eighty-Four Minutes, Ninety-Four Lives*. Worcester, MA: Data Books, 1993.

Louise Quayle, *Weather: Understanding the Forces of Nature*. New York: Crescent Books, 1990.

Periodicals

Wallace Akin, "The Great Tri-State Tornado," *American Heritage*, May/June 2000.

Bill Briggs, "The Storm Tornado Chasers Play Perilous, Costly Game," *Denver Post*, June 1996.

Meg Jones, "Predicting Severity, Location of Tornadoes Is Difficult, Experts Say," *Milwaukee Journal Sentinel*, April 2001.

David M. Ludlum, "Most U.S. Tornado Deaths in a Single Day," *Weatherwise*, January/February 1998.

Betsy Mason, "Scientists Seek Better Forecasting by Targeting 'Chaos Hot Spots,'" *Dallas Morning News*, July 2001.

Dennis McCown, "Close Encounter with a Rocky Mountain Whirlwind," *Weatherwise*, June/July 1997.

James R. McDonald, "T. Theodore Fujita: His Contribution to Tornado Knowledge Through Damage Documentation and the Fujita Scale," *Bulletin of the American Meteorological Society*, January 2001.

Daniel Pendrick, "Tornado Troopers," *Earth*, October 1995.

Joan Phillips, "Journal: Tornado Relief Center," *Families, Systems, and Health: The Journal of Collaborative Family HealthCare*, Winter 2000.

W.T. Roach and J. Findlater, "An Aircraft Encounter with a Tornado," *Meteorological Magazine*, February 1983.

C.J. Root, "Some Outstanding Tornadoes," *Monthly Weather Review*, 1926.

Jeff Rosenfeld, "Mr. Tornado," *Weatherwise*, May/June 1999.

Andrew J. Sowder, "Tornado Warning!" *News Photographer*, May 1998.

Cynthia Ramsay Taylor, "Building Disaster-Resistant Communities," *USA Today*, July 2001.

Linda Wang, "Seismic Shivers Tell of Tornado Touchdown," *Science News*, February 2001.

Cecilia Wessner and Karoline Patter, "What Is the Difference Between a Cyclone and a Tornado?" *Popular Science*, February 2000.

Jeff Wise, "Playing Twister," *Women's Sports & Fitness*, June 2000.

Websites

Federal Emergency Management Agency (FEMA), www.fema. gov. FEMA has tornado safety tips brochures and a tornado fact sheet.

National Climatic Data Center, www4.ncdc.noaa.gov. The center's website has details on all known tornadoes since 1993 and is updated monthly.

National Oceanic and Atmospheric Administration (NOAA), www.noaa.gov. NOAA gives immediate access to all available tornado warnings for the United States as well as general tornado information.

National Severe Storms Laboratory, www.nssl.noaa.gov. This research lab is internationally known and gives the latest tornado research information.

National Weather Service, http://iwin.nws.noaa.gov. The National Weather Service is an interactive weather information network that provides all national weather service home pages as well as live computer data broadcasts via satellite, radio, and the Internet.

Storm Prediction Center, www.spc.noaa.gov. The center's website has a summary of all severe weather of the previous day and official statistics for the past several years.

Tornadoes and Tornado Research, www.geocities.com/joefurr2. This site provides information on tornado formation, safety, statistics, research, historical tornadoes, tornado pictures, and remote sensing.

Tornadoes.com, www.tornadoes.com. This is a comprehensive resource guide to tornadoes with a list of recommended tornado books.

Tornadoes Links, www.sirlinksalot.net. This site will direct you to the latest tornado news and the best tornado websites.

Tornadoes Theme Page, www.cln.org. This site is a collection of useful Internet educational resources to supplement the study of tornadoes.

Tornado Project, www.tornadoproject.com. This website specializes in tornado myth, tornado oddities, personal experiences, tornado chasing, safety, and tornadoes from the past. It has updated statistics and World Wide Web links.

Weather Channel, www.weather.com. The Weather Channel allows you to find your local weather forecast and provides weather-related videos for viewing.

INDEX

DATE DUE

MAR 2 2 2010		MAR 2 3 2010	
		JUL 0 1 2010	
		JUN 2 8 2010	
OCT 1 8 2010		NOV 0 1 2010	
NOV 1 5 2010		NOV 0 1 2010	
		NOV 0 3 2010	
FEB 0 5 2014		FEB 0 5 2014	